Green Research, Developments and Programs

nova science publishers

www.novapublishers.com

Green Research, Developments and Programs

Sustainable and Healthy Building Environments
Roberto Alonso González Lezcano, PhD (Editor)
2023. ISBN: 979-8-88697-401-0 (Hardcover)
2023. ISBN: 979-8-88697-497-3 (eBook)

Green Technology for Bioremediation of Environmental Pollution
Bana Bihari Jana, Jatindra Nath Bhakta
and Susmita Lahiri (Editors)
2019. ISBN: 978-1-53614-528-1 (Hardcover)
2019. ISBN: 978-1-53614-529-8 (eBook)

The Power-Saving Behavior of Households: How Should We Encourage Power Saving?
Kenichi Mizobuchi and Hisashi Tanizaki (Editors)
2018. ISBN: 978-1-53613-173-4 (Softcover)
2018. ISBN: 978-1-53613-423-0 (eBook)

Greenhouse Gas Emissions and Nitrogen Losses from Grazed Dairy and Animal Housing Systems
Jiafa Luo and Yan Li (Editors)
2017. ISBN: 978-1-53611-100-2 (Hardcover)
2017. ISBN: 978-1-53611-100-2 (eBook)

Deciphering Organic Foods: A Comprehensive Guide to Organic Food Production, Consumption, and Promotion
Darrel D. Muehling and Ioannis Kareklas (Editors)
2017. ISBN: 978-1-53610-517-9 (Hardcover)
2017. ISBN: 978-1-53610-524-7 (eBook)

More information about this series can be found at
https://novapublishers.com/product-category/series/green-research-developments-and-programs/

Jacques Kent
Editor

The Future of Green Synthesis

nova science publishers

www.novapublishers.com

DOI: https://doi.org/10.52305/YIWU6826

NOTICE TO THE READER

Library of Congress Cataloging-in-Publication Data

ISBN: 979-8-88697-816-2

Published by Nova Science Publishers, Inc. † New York

Contents

Preface

Currently, the synthesis process is one that is associated with toxic, non-biodegradable chemicals that are not safe for our environment. This book focuses on the trends, recent advancements and future of green synthesis. It discusses the advantages green synthesis has in terms of creating a more environmentally friendly, non-toxic, and less time-consuming method of synthesis.

Chapter 1 - Although fire is no longer commonly used in synthetic chemistry, Robert Bunsen's invention of the burner in 1855 allowed the energy of this heat source to be focused, especially on a reaction vessel. Later, the Bunsen burner was replaced by the isomantle, oil bath, or hot plate as a source for supplying heat to a chemical reaction. A novel heating technique that makes use of microwaves is known as the "Bunsen burner of the twenty-first century." It is gaining popularity for reducing reaction time dramatically (from days and hours to minutes and seconds). The use of microwave energy to heat and speed up chemical reactions has grown among scientists in recent years. The chemical transformations are extremely effective since they have greater yields, reactivity, and selectivity. Microwave irradiation is regarded as an important component of green chemistry because it generates clean compounds with no residual toxins. The process of producing chemical components in a sustainable manner is recognized as "Green Chemistry." Green chemistry aims to reduce the harmful impact that the production and manufacturing of various chemical compounds have on the environment. Green chemistry seeks to produce compounds with the fewest adverse components by improving chemical synthesis efficiency and generating as little trash as possible. The future of green chemistry includes the use of many potent techniques, such as the microwave irradiation approach. Chemical research and production have undergone a revolution with the use of microwave assisted methods. The use of cutting-edge microwave-assisted chemistry techniques combined with environmentally friendly reaction media is substantially reducing hazardous waste and reaction times in a number of organic synthesises. An overview of the authors' specific efforts in generating MW-assisted organic transformations is given for the synthesis of several heterocyclic compounds, coupling reactions, oxidation-reduction processes, and certain named reactions. These transformations make use of a number of safe substitutes, including solid-supported reagents and more environmentally friendly reaction media, particularly aqueous, ionic liquid, deep eutectic solvents, solvent-free, and catalyst-free systems. This chapter enlightens us about the principles and advantages of microwave heating. In addition, it includes microwave-assisted organic synthesis with green or without reaction media.

Chapter 2 - Different unique properties of zinc oxide (ZnO) nanomaterials allow it to occupy a prominent position in modern technologies in various aspects. This chapter covers the green route synthesis of ZnO nanomaterials and details the future prospects of these nanomaterials. The green synthesis method has many advantages over the traditional physical and chemical synthesis methods in terms of more manageable, less time-consuming, inexpensive, non-toxic, environmentally friendly, reliable and sustainable development of modern society. Green synthesis belongs to the 'bottom-up' approach of nano-synthesis processes, similar to chemical reduction, in which reducing agents are replacing by plant extracts (containing the phytochemicals like polyphenols, flavonoids and terpenoids etc.) having inherent stabilizing and capping properties. During the green synthesis process, plant extracts are mixed in a zinc precursor solution. The precursors act as the source of zinc ions. Finally, the ions are reduced by the phytochemicals present in the extract and precipitates as zinc oxide nanomaterials. A series of experimental techniques are implemented for the usual confirmatory testing of the synthesized materials and results show that green synthesized zinc oxide nanomaterials can exhibit excellent antibacterial, anticancer, antioxidant and photocatalytic dye degradation etc. properties. In areas where purity is the primary concern, such as for example, in medical, biological, agricultural and food industries, the development of plant-based ZnO nanomaterials free from chemical impurities is required. From this point of view, detailed study is required to know the green mechanisms related to the principles of green chemistry and their activities during the plant-based synthesis of ZnO nanomaterials.

Chapter 3 - Silver nanoparticles (AgNPs) are being widely studied due to their physical and chemical properties, which depend mainly on morphology, size, and composition, where the stabilizing agent is of great importance (Donahue et al. 2022).

When talking about obtaining AgNPs, several synthesis protocols are established, which include chemical (electrochemical, photochemical, and chemical reductions, among others) and physical methods (mainly laser ablation).(Iravani et al. 2014)However, derived from the interest in giving biomedical applications to AgNPs (antimicrobial and antioxidant activities or biosensing), a series of new methodologies have emerged called « green synthesis methods,» where biopolymers, biomass, plant extracts, and other varieties of natural products are used. Green synthesis methods can produce a new generation of AgNPs with low toxicity and cost (Oceguera-Galindo, Oceguera-Contreras, and Pozas-Zepeda 2020).

The main objective of this review is to present the information derived from various investigations where plant extracts function as reducing agents for AgNPs, emphasizing the phytochemical composition of natural plant extracts. Biomolecules, naturally found in plant extracts, are responsible for reducing the metal ions in solution forming the AgNPs, and subsequently, some compounds can stabilize the AgNPs. Among the biomolecules studied with reducing capacity are carbohydrates, phenolic acids, flavonoids, proteins, and free amino acids. These molecules possess specific chemical characteristics that make them essential for understanding the mechanisms of AgNPs formation. Likewise, the pros and cons of the various synthetic methodologies and the biomedical applications presented by this new green generation of AgNPs are widely discussed.

Chapter 4 - Developing useful products with biotechnological intervention has recently gained wide applicability and acceptability. Green chemistry and enzyme-mediated biotransformation are two aspects that have witnessed novel advancements having far-reaching implications. The amalgamation of applied microbiology, biotechnology, genetic engineering, immobilization, bioprocess development and downstream processing has inherently altered every facet of product synthesis. This chapter's main goal is to showcase microbial enzymes' role in green synthesis. The synthesis of useful products in various industries like medicine, chemical and agriculture is being pursued through enzyme-catalyzed processes. The main reason for this trend is the eco-friendly nature of enzymatic technologies required for attaining the twin goals of economic growth and sustainable development. Enzymes display multiple unique attributes making them highly indispensable for future industrial endeavors. Green chemistry aims to reduce the addition of toxic compounds into the environment, and one of the key technologies in realizing this objective is biocatalysis. With increasing emphasis on scale-up and operational efficiency, key enzyme-based technologies have evolved and played a significant role in industrial production.

Chapter 5 - Patients with immune compromise, such as those infected with HIV, organ transplant recipients, chemotherapeutic patients, and patients requiring long-term antimicrobial treatment, are increasingly difficult to treat. Hence, a comparatively new and remarkable class of antifungal drugs has to be developed in a cost-effective green synthesis pathway to solve this important medical problem. Since the last three years, from 2019 to 2022, the number of COVID-19 patients throughout the globe has been increasing, and its treatment is still a problem. No special drugs are produced for its treatment. In the last three decades, the diseases associated with invasive fungal infections have gradually increased throughout the world, and their proper treatment is a problem for medical science. In this regard, it was reported that some specific organic compounds, such as fluconazole, itraconazole, posaconazole, voriconazole, and ravuconazole, were found to be more influential in comparison with other antifungal drugs. The present article focuses on explaining the various methods of green synthetic approaches to Fluconazole, Itraconazole, Posaconazole, Voriconazole, and Ravuconazole using cost-effective and sustainable approaches.

Chapter 6 - Many research groups around the world have been dealing with the synthesis of dronic acids and their derivatives for decades. These compounds are extremely effective in the treatment of osteoporosis and bone tumors. Their advantage is that they can be used in small doses, but due to their difficult preparation and complicated cleaning, their prices are very high. The industry's most commonly used solution for the synthesis of α-hydroxy-methylenebisphosphonates is the so-called "direct" synthesis, in which case the corresponding amino acids, heteroaryl acetic acids or derivatives are reacted with various P-reagents. The advantage of the method is that the reagents are easily accessible and relatively cheap, but their excessive use pollutes the environment. The disadvantages of this method are the extremely heterogenous reaction of the mixture, viscosity, difficulties in mixing and the released large amount of hydrochloric acid gas by the hydrolysis. These problems are manageable in laboratory environment, but it is difficult to perform these transformations in the industrial environment and to keep these reactions in safe conditions. The "indirect" synthesis may also prepare the valuable drugs through α-oxophosphonates utilizing the Pudovik reaction. There are other possibilities for the preparation of the active ingredients. Some of them are presented in this chapter.

Chapter 7 - To maintain sustainable development of our globe, organic materials should be synthesized by using the twelve principles of green chemistry. Due to the generation of different variants of COVID-19 (Corona virus) from time to time, the need for anti-COVID-19 drugs is vastly increasing. There is a great challenge in this situation to meet this challenge. The authors must develop a methodology of preparing anti-COVID-19 drugs based on eco-friendly processes that should satisfy pharmaceutical industries. Today, most of the synthesis process proceeds with several reactions associated with numerous toxic chemicals in the form of substrates, reagents, catalysts, additives, and solvents. Various challenges were faced during the storage, quenching, and transportation of toxic chemical substances. Since most of the chemicals used are not biodegradable, it is a serious threat to our ecosystem and biological community of the globe. Hence, to solve this global crisis, numerous green technologies, such as less hazardous chemical synthesis process must be adopted, leading to the use of bio-resources and water as solvents, one-pot synthesis, etc. The method of one-pot synthesis is a chemical reaction that is carried out in a single reactor, and no purification of the intermediate products is required. For organic intermediate purification, huge quantities of toxic organic solvents, such as chlorinated solvents including CH2Cl2, CHCl3, and CCl4 and highly flammable solvents such as hexane and benzene, are required. Hence, one-pot synthesis is not only a green pathway of synthesis but is also cost-effective. Therefore, to minimize the pollution level in the environment, various green technology pathways should be developed for the synthesis of drugs and organic compounds. The present chapter summarizes remarkable green chemistry initiatives in anti-COVID-19 drugs based on organic molecules. The drugs used for COVID-19 treatment are highly effective for treatment of viral diseases that primarily affect the respiratory system, where an epithelial cell of the respiratory system is damaged, resulting in mucosal membrane inflammation. Inflammatory mucosal membranes lead to alveolar damage and eventually pneumonia. Covid-19 is caused by a positive-sense single-stranded RNA virus, SARS-CoV-2, usually known as the novel corona virus. Even though many drugs have been screened and tested in clinical trials, only a few of them, such as hydroxychloroquine, favipiravir, remdesivir, molnupiravir, etc., have attracted scientists and shown remarkable activity against SARS-CoV-2.

Chapter 1

Microwave Assisted Synthesis: A Green Chemistry Approach and Future Directions

S. Makwana
and P. Kumari*
Department of Chemistry, Sardar Vallabhbhai National Institute of Technology, Gujarat, India

Abstract

Although fire is no longer commonly used in synthetic chemistry, Robert Bunsen's invention of the burner in 1855 allowed the energy of this heat source to be focused, especially on a reaction vessel. Later, the Bunsen burner was replaced by the isomantle, oil bath, or hot plate as a source for supplying heat to a chemical reaction. A novel heating technique that makes use of microwaves is known as the "Bunsen burner of the twenty-first century." It is gaining popularity for reducing reaction time dramatically (from days and hours to minutes and seconds). The use of microwave energy to heat and speed up chemical reactions has grown among scientists in recent years. The chemical transformations are extremely effective since they have greater yields, reactivity, and selectivity. Microwave irradiation is regarded as an important component of green chemistry because it generates clean compounds with no residual toxins. The process of producing chemical components in a sustainable manner is recognized as "Green Chemistry." Green chemistry aims to reduce the harmful impact that the production and manufacturing of various chemical compounds have on the environment. Green chemistry seeks to produce compounds with the fewest adverse components by improving chemical synthesis efficiency and generating

* Corresponding Author's Email: pl@chem.svnit.ac.in.

In: The Future of Green Synthesis
Editor: Jacques Kent
ISBN: 979-8-88697-816-2

as little trash as possible. The future of green chemistry includes the use of many potent techniques, such as the microwave irradiation approach. Chemical research and production have undergone a revolution with the use of microwave assisted methods. The use of cutting-edge microwave-assisted chemistry techniques combined with environmentally friendly reaction media is substantially reducing hazardous waste and reaction times in a number of organic synthesises. An overview of our specific efforts in generating MW-assisted organic transformations is given for the synthesis of several heterocyclic compounds, coupling reactions, oxidation-reduction processes, and certain named reactions. These transformations make use of a number of safe substitutes, including solid-supported reagents and more environmentally friendly reaction media, particularly aqueous, ionic liquid, deep eutectic solvents, solvent-free, and catalyst-free systems. This chapter enlightens us about the principles and advantages of microwave heating. In addition, it includes microwave-assisted organic synthesis with green or without reaction media.

Keywords: microwave irradiation, dielectric heating, green chemistry, organic synthesis, applications

Introduction

The goal of green chemistry is to reduce the rising toxicity levels in our environment. It is a way of thinking and a project that advocates for and carries out eliminating the use of hazardous materials that would otherwise be used in customary synthetic chemical processes. Sustainable development was defined by the Brundtland Commission (1987) as "development that meets the requirements of the present without compromising the ability of future generations to meet their own needs." We would be successful in our efforts to guide the earth toward a sustainable future if we maintain the fundamentals of green chemistry in mind while carrying out the typical organic transformations, along with significant research (L. Bernard, A. Friis-christensen 2008). According to the 12 principles of green chemistry, which result in sustainability in a chemical world in terms of developing sustainable sources of feed stocks and resources, eliminating the usage of undesirable substitutions and alternates, preventing waste chemicals, increasing atomeconomy, using less hazardous chemical syntheses, incorporating safer solvents and auxiliaries, and designing for better energy efficiency (Laskar et al. 2022).

Green technology helps to speed up processes and their improvements in the R&D sector by being introduced to the drug development industries (Belwal 2014). The R&D mindset fits with green chemistry concepts is illustrated in Table 1 (Rogers and Jensen 2019).

Table 1. Green chemistry is cost-efficient (Rogers and Jensen 2019)

	Environmentally Thinking	Economically Thinking
Atom Economy	Minimal by-product formation	More from less - incorporate the total value of materials
Solvent Reduction	Less solvent waste	Higher throughput, less energy
Reagent Optimization	Catalytic, low stoichiometry, recyclable reagents minimize usage	Higher efficiency - higher selectivities
Convergence	due to increased process efficiency	Higher efficiency - fewer operations
Energy Reduction	from power generation, transport, and use	Reduced energy reflects increased efficiency, shorter process, mild conditions
In-situ Analysis	Reduced possibility for exposure or release to the environment	Real-time data increases throughput and process efficiency, with fewer reworks
Safety	Non-hazardous materials reduce the risk of exposure, release, explosions &fires	Worker safety and reduced downtime Reduced time on special control measures

Evolution of Microwave Irradiation and Its Application in Green Chemistry

Green chemistry applies a set of principles to the design, production, and use of chemical products associated with a particular synthesis or process to minimize or fully prevent the use or production of hazardous substances. Chemists can significantly lower the risk to both the environment and human health by utilizing these principles. Green Chemistry is not a brand-new field of study but it is a nobel philosophical perspective that could advance sustainable development through the use and expansion of its guiding principles (de Marco et al. 2019) (Figure 1). It is simple to locate numerous fascinating examples of how to apply green chemistry principles., e.g., Check or avoid overproduction of bio-accumulative, bio-transforming, non-biodegradable substances such as polythene, Aldrin, Chlordane, DDT, and methyl mercury compounds. They help to carry out cShemical reactions and assess environmental impact.

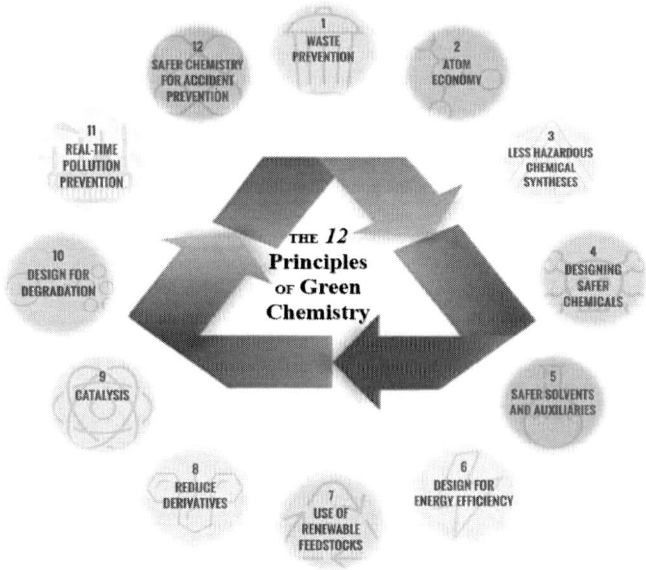

Figure 1. Principles of green chemistry (de Marco et al. 2019).

There are several guides on the toxicity of solvents, which many institutions and companies carry out. However, Prat and collaborators made an interesting comparison of several of these guides reported in 2014 and compiled in a table, in which they divided the solvents into six categories: "recommended", "recommended or problematic?", "problematic", "problematic or hazardous?", "hazardous" and "highly hazardous (Prat, Hayler, and Wells 2014) (Table 2). This information could be beneficial in designing a synthetic protocol.

These principles educate us that chemistry ought to be effective, non-wasteful, and non-hazardous rather than producing undesirable and toxic materials that require energy to deal with later. A design framework for innovative chemical technologies is provided by these guiding principles. Chemists should embrace each of these concepts and put them into practice. The green philosophy can and is used in analytical chemistry and polymer research, but they are particularly well suited to the synthetic community. Green chemistry can be thought of as a series of reductions (European Commission. 2001) (Figure 2). The goal of improving social, environmental, and economic well-being is reinforced by these cuts.

Table 2. Solvent classification
(de Marco et al. 2019; Prat Hayler, and Wells 2014)

Recommended	• Water • Ethanol • Isopropyl alcohol • N-butanol • Ethyl acetate	• Isopropyl acetate • Butyl acetate • Anisole • Sulfolane	-Toxic
Recommended or Problematic?	• Methanol • Tert-butanol • Benzyl alcohol • Ethylene glycol • Acetone • Butanone	• Methyl isobutyl • Ketone • Cyclohexanone • Methyl acetate • Acetic acid • Acetic anhydride	
Problematic	• 2-Methyl tetrahydrofuran • Heptane • Me-cyclohexane • Toluene	• Xylenes • Chlorobenzene • Acetonitrile • Dimethyl-propyleneurea • Dimethyl sulfoxide	
Problematic or Hazardous?	• Methyl tert-butyl ether • Tetrahydrofuran • Cyclohexane	• Dichloromethane • Formic acid • Pyridine	
Hazardous	• Di-isopropyl ether • 1,4-dioxane • Dimethoxyethene • Pentane • Hexane	• N, N'-dimethyl formamide • N, N'-dimethyl Acetamide • Methylpyrrolidone • Methoxy-ethanol • Triethanolamine	
Highly Hazardous	• Diethyl ether • Benzene • Chloroform	• Carbon tetrachloride • Dichloroethane • Nitromethane	+Toxic

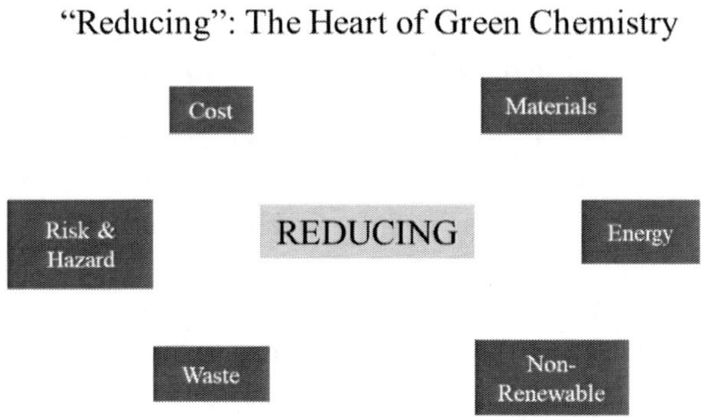

Figure 2. "Reducing": the heart of green chemistry (Bassyouni, Abu-Bakr, and Rehim 2012).

Current synthetic technologies are increasingly being overtaken by longer-lasting, environmentally and mentally friendly methods. There is a range of technologies that are increasingly being used in an attempt to realize the objectives of green chemistry. The key "clean technologies" are listed in (Bassyouni, Abu-Bakr, and Rehim 2012) (Figure 3). As energy costs escalate and more needs to be done to minimize the emissions associated with energy production, energy consumption becomes more significant in green chemistry. This paves the way for the utilization of alternate energy sources and better-designed reactors. In this respect, to use intensely focused radiation, the potential of microwave reactors to produce reactions with much faster reaction rates, significantly greater product yields, and even more, selective reactions are exploited.

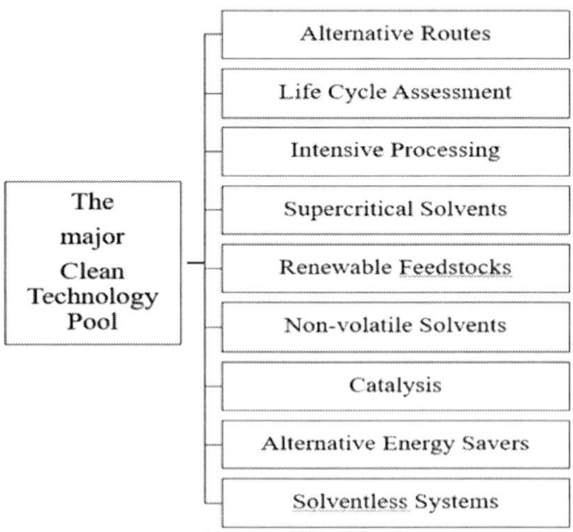

Figure 3. The major clean technologies (Bassyouni, Abu-Bakr, and Rehim 2012).

Anastas defined and emphasized the twelve generally acknowledged tenets of green chemistry. (Erythropel et al. 2018) The fifth, sixth and ninth principles may initially appear to have a direct connection to microwave chemistry upon evaluation of these concepts. Particularly, the application of solvent-free conditions (5th principle) or environmentally friendly solvents (such as water and ionic liquids) in combination with microwave heating techniques has frequently been done, and generally speaking, these conditions have been promoted as being environmentally friendly (Polshettiwar and Varma 2008). Catalytic transformations (9th principle are quite common in microwave chemistry (Desai and Kappe 2004).

These microwave-assisted transformations may very well be green, the relative greenness of these processes is not always related to the use of microwave energy but rather is a result of the environmentally friendly reaction media. The greenness of the microwave heating technique itself needs to be assessed. A careful evaluation of the energy efficiency of microwave dielectric heating technology is necessary for front of the twelve principles of green chemistry (6th principle) (Anastas and Kirchhoff 2002).

Microwaves in Green Synthesis

Microwaves weren't used in the synthesis of organic compounds until the middle of the 1980s. The research teams of Richard Gedye (Gedye et al. 1986) and Camelia Gabriel (Giguere et al. 1986) have published the first scholarly studies on the use of microwave heating to mediate organic processes. These early microwave-assisted organic synthesis (MWAOS) studies were frequently conducted in sealed Teflon or glass pots in a domestic microwave oven without any temperature or pressure reading controls. In those early days, there were several explosions caused by the quick, uncontrolled heating of organic solvents, but as time went on, more scientists began to study this new technology. The first attempts at dry-media microwave chemistry (which reduced the risk of explosions) were made in the 1990s (de la Hoz, Díaz-Ortiz, and Prieto 2015; Bassyouni, Abu-Bakr, and Rehim 2012; Sakai, Hirayama, and Tamura 2006; Varma 2001; Varma 2002; Varma 1998; Loupy et al. 1998). The solvent-free technique was particularly popular since it allowed the safe application of conventional open-vessel technology and domestic microwave ovens. This technique reduces the amount of leftover waste solvent produced. In addition, such reactions also typically result in products that don't need much purification.

The method developed by Christopher R. Strauss in the mid-1990s (Strauss 2004; Strauss and Trainor 1995) for MWAOS in dedicated sealed vessels utilizing common solvents has also recently regained popularity among chemists. This methodology will probably become the standard procedure for carrying out MWAOS on a laboratory scale in the future because of the advantageous combination of quick heating by microwaves with sealed-vessel (autoclave) technology. Presently, "microwave chemistry" has emerged as a disguised boon for environmentally conscientious scientists. The organic reactions (Sanghi 2000) mediated by microwaves occur more quickly, securely, and sustainably, with a higher yield. One of the main benefits of microwave chemistry is that it uses a very little solvent, water as a solvent, or even a dry medium on a solid support.

Conventional Reflux to Microwave Heating

Chemical laboratories frequently use reflux systems for traditional heating, where the reaction temperature is typically controlled by the solvent's boiling point. Since this is an open system that cannot be compressed, or superheated to achieve greater temperatures.

Due to these factors, traditional chemical reactions can take a very long period. It is not unusual for chemical synthesis to take many hours. Often, it takes a reaction all night or even all weekend to complete and convert all of the starting materials.

The Arrhenius law is the key to speeding these processes. According to this rule of thumb, raising the temperature by 10°C twice the rate of a reaction, allowing it to be completed considerably faster i.e., a temperature increase of only 20°C causes the reaction time to decrease to one-fourth of the initial value.

Example

A reaction that is performed in boiling *ethanol* (approx. 80°C) within 8 hours can be performed at 160°C within only 2 min.

All synthesis and sample preparation reactors that are commercially available operate under the same general principle, reaction vessels are sealed and closed before being heated above the boiling point of the reaction mixture. Due to the utilization of extremely high temperatures, the reaction time is shortened and reaction products are produced in a matter of minutes as opposed to hours or even days.

Pharmaceuticals, polymers, cosmetics, nanomaterials, agrochemicals, and other useful products are produced through chemical synthesis. Since time truly equals money, the chemical industries have a keen interest in increasing productivity and efficiency in scale-up.

The most time-consuming step in the traditional synthesis is usually optimization, which involves determining the best conditions for a given reaction to produce the desired products with good yields and purity. These optimizations are frequently challenging and time-consuming since many synthesis reactions need at least one heating step and/or several heating stages that last for a lengthy time. Since it frequently drastically minimizes reaction times typically from days or hours to minutes or even seconds microwave-assisted heating under controlled circumstances has been proven to be an invaluable technology for any application that requires heating a reaction mixture. Therefore, this promising method enables the rapid synthesis of compounds in either parallel or (automated) sequential manner.

As a result of how quickly different reaction conditions can be tested, a morning conversation of "What should we try?" can quickly turn into a discussion of "What were the results?" after lunch. Since microwave synthesis is already a key methodology in most pharmaceutical, agrochemical, biotechnology, and material science businesses' chemical laboratories, it should come as no surprise. They've come to understand how microwave technology may speed up chemical reactions, and by transformation, their entire production process (Kappe et al. 2012).

> *"Microwaves do have the potential to become the Bunsen burners of the 21st century."*
>
> Prof. C. Oliver Kappe, University of Graz.

Is Your Synthetic Protocol Green?

For a medicinal chemist, this significant question needs to be addressed. It is possible to evaluate the greenness of a certain synthesis procedure using the twelve green chemistry principles. These guidelines address a number of issues, including the use of different solvents and the volume of chemical waste produced, the quantity and reusability of catalysts and reagents, the number of chemical steps (energy efficiency and atom economy), the use of safer chemicals, and reaction conditions (Valavanidis 2016; Kappe et al.; Anastas and Kirchhoff 2002) (Figure 4). A new synthetic protocol following all 12 principles is very challenging and not expected. However, the more principles a protocol can satisfy, the greener the developed process will be.

Because of the size and complexity of this area, there are many different approaches utilized to develop sustainable methods. Making a protocol more environmentally friendly involves using alternative reaction media. However, replacing the solvent alone might not be sufficient. The solvent is simply one piece of this puzzle, which requires careful attention throughout the entire procedure. It's also important to consider energy usage, atom efficiency, and the utilization of renewable resources.

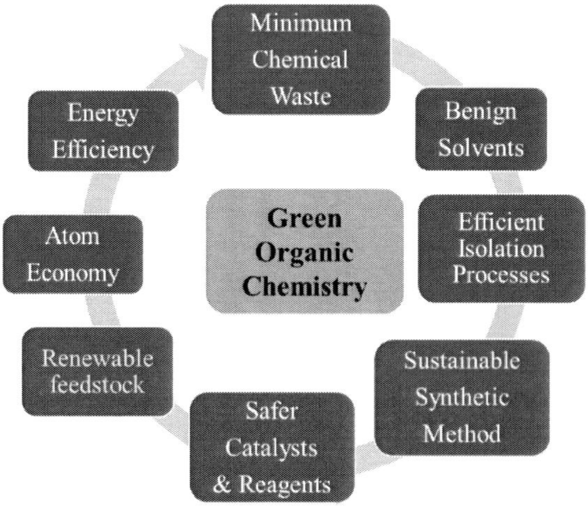

Figure 4. Principle of green and sustainable organic chemistry (Valavanidis 2016).

Solvents play a significant role in how environmentally friendly chemical industry processes are, as well as how expensive, safe, and healthy they are. They are a major focus of green chemistry since they are very combustible and volatile, the main cause of environmental contamination The use of solvents in reactions cannot be avoided since they are required for several processes, such as reactant mixing, a steady and uniform supply of energy through heat transfer, and in some instances, regulation of the regio- and chemo-selectivity of reactions (Clark and Tavener 2007). However, the development of atom-economic synthetic techniques, which selectively generate the desired product without producing any by-products, can prevent or minimize the use of organic solvents for the isolation and purification of products.

Environmental improvements connected to solvents can be achieved by putting into practice several distinct strategies;

(1) Substituting safer, healthier, and more environmentally friendly solvents for dangerous ones.
(2) Bio-solvents: These are solvents made from renewable resources, including ethanol made from the fermentation of feeds with sugar or starchy feeds.

(3) Substituting environmentally friendly supercritical fluids like CO_2 and safe ionic liquids for organic solvents. These fluids have low vapor pressures, which prevent environmental emissions.
(4) Biphasic technologies: which combine supercritical carbon dioxide, aqueous systems, and fluorous and regenerable ionic liquids.

Principles of Microwave Heating

The electromagnetic radiation known as "microwave irradiation" has a frequency from 0.3 to 300 GHz and a wavelength of between 1 mm and 1 m. (Figure 5) (Kumar et al. 2020). The frequency of operation for all household "kitchen" microwave ovens and commercially available microwave reactors specifically designed for chemical synthesis is 2.45 GHz (corresponding to a wavelength of 12.25 cm)

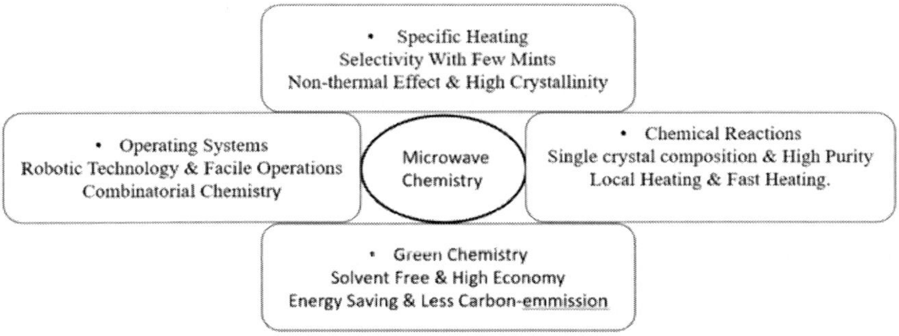

Figure 5. The representation of multidimensional advantages of microwave chemistry with an appropriate combination of (i) operating systems, (ii) green chemistry approach, (iii) specific heating, and (iv) chemical reactions. MW, microwave. (Kumar et al. 2020).

The energy of the microwave photon at a frequency of 2.45 GHz (about 10^{-5} eV) is too low to sever molecular bonds and is also lower than Brownian motion (Pal and Gayen 2021; Nüchter et al. 2004) (Table 3). Thus, it is evident that, in contrast to UV and visible radiation, microwaves cannot "induce" chemical reactions through the direct absorption of electromagnetic energy (photochemistry).

Table 3. Comparison of radiation types and bond energies, frequency (Pal and Gayen 2021; Nüchter et al. 2004)

Radiation type	Frequency (MHz)	Quantum energy (eV)	Bond type	Bond energy (eV)
Gamma rays	3.0×10^{14}	1.24×10^{6}	C-C single bond	3.61
X-Rays	3.0×10^{13}	1.24×10^{5}	C=C double bond	6.35
Ultraviolet	1.0×10^{9}	4.1	C-O single bond	3.74
Visible light	6.0×10^{8}	2.5	C=O double bond	7.71
Infrared light	3.0×10^{6}	0.012	C-H bond	4.28
Microwaves	2450	10^{-5}	O-H bond	4.80
Radio frequencies	1	4.0×10^{-9}	Hydrogen bond	0.04-0.44

On the other hand, microwave irradiation produces distinctive thermal effects that are very helpful for chemical synthesis. Only higher energy radiation (such as UV or visible light, which causes photochemistry) can break molecular bonds, which therefore induces chemical processes.

Microwave Dielectric Heating

Microwave chemistry operates based on the adequate heating of materials (often solvents) by dielectric heating effects. Two main mechanisms drive microwave dielectric heating (Gabriel et al. 1998) are:

- Dipolar polarization

  When exposed to microwave radiation, a substance needs to have a dipole moment to produce heat. The dipoles in the field line up with the microwave field's oscillations. This alignment causes rotation, which in turn causes friction, which generates heat energy (Figure 6).
- Ionic conduction

  Ions that are dispersed in the solvent bounce back and forth as a result of microwave radiation, i.e., ionic conduction. The charged particles collide with surrounding molecules or atoms as a result of this oscillation, generating heat energy. For instance, distilled water and tap water can be heated to the same temperature using microwave radiation, but tap water will heat up more quickly because of its ionic content and the dipolar rotation of the water molecules (Mgbemena et al. 2018) (Figure 6).

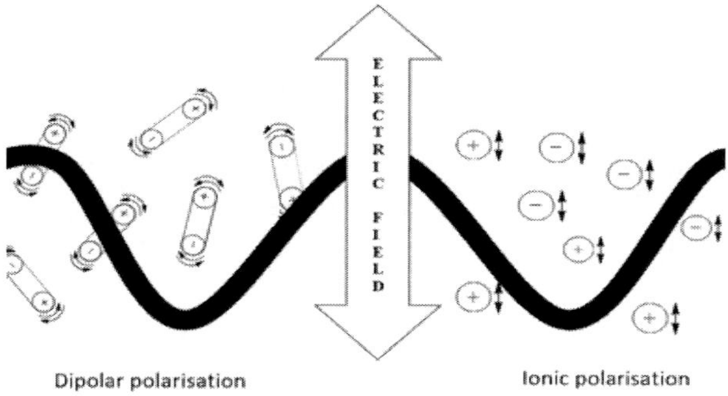

Figure 6. Schematic illustration of the two main dielectric heating mechanisms: dipolar polarization (dipoles align in the microwave field) and ionic conduction (ions move in the microwave field) (Mgbemena et al. 2018).

Microwave heating of gases and solid is hardly possible! Because of the too huge gap between the revolving molecules, gases cannot be heated by microwave radiation. Similar to gases, solids like ice are almost transparent to microwave radiation because the water dipoles cannot move as freely as they may when they are liquid since they are trapped in the crystal lattice. However, some conductive solid materials, such as silicon carbide, in which electrons can freely travel, are efficient microwave absorbers and hence heat up very quickly.

With dielectric heating, electric energy is converted into kinetic energy which is ultimately converted into heat.

Microwave Heating Over Conventional Heating

The conventional method of doing organic synthesis involves refluxing a reaction mixture utilizing a heated oil bath as a heat source. However, the effectiveness and efficiency of this approach to heating a reaction mixture are reduced since heat energy must be transferred from the hot oil bath to the reaction vessel's surface before the hot surface can heat the reaction vessel's contents (Gabano and Ravera 2022) (Figure 7). The hot surface can also cause localized overheating and the degradation of delicate materials.

In contrast, microwave irradiation produces energy-efficient internal heating through a direct connection of microwave energy with the reaction mixture's ions and dipoles. The reaction mixture is heated on a molecular level by direct interaction with the molecules when microwaves travel through the (nearly) microwave-transparent vessel wall (solvents, reagents, catalysts, etc.., absorption). As opposed to a system that is heated traditionally, microwave irradiation produces inverted temperature gradients since it heats the system directly "in-core" (without first heating the vessel surface) (Gabano and Ravera 2022) (Figure 7).

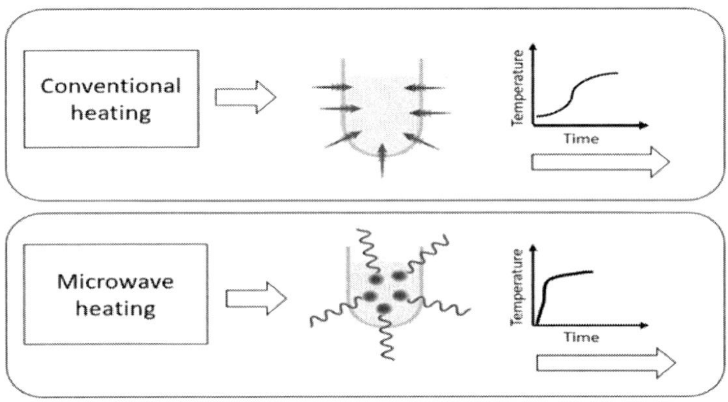

Figure 7. Graphical illustration of heating introduction and temperature distribution into a reaction mixture for conventional heating and microwave heating (Gabano and Ravera 2022).

Modern microwave reactors use reaction vessels typically made of (almost) microwave-transparent materials like borosilicate glass, quartz, or Teflon which allows radiation to pass through the vessel's walls and produce an inverted temperature gradient compared to traditional thermal heating. The sample will see a homogenous temperature increase if the microwave cavity is well-designed. Because of the highly effective internal heat transfer, there are fewer wall effects (no hot vessel surface).

Additionally, the conversion of electromagnetic energy into heat energy is exceedingly effective and produces heating rates that are incompatible with conventional heating. The generation of by-products is reducing because the target temperature is quickly reaching. This is another fantastic benefit of microwave since it allows for greater product yields and streamlines the process.

Benefits of Microwave Synthesis in Dedicated Microwave Reactors

⇒ Possibility of convenient superheating of solvents.

- achieve higher temperatures.
- dramatically reduce your overall process time.
- obtain higher yields and purer compounds.

⇒ Superior parameter control.

- increased repeatability.
- maximum handling comfort.

⇒ Access to automated setups and parallel synthesis

- increase your efficiency.
- time equals money!

⇒ Potential for stirring.

- improve the temperature homogeneity.
- strengthen the reliability of your findings in the scientific community.

⇒ Continuous power output.

- complex reaction management.
- safe and efficient processing.

⇒ Touchscreen interface with an intuitive user interface.

- convenient programming and data management.
- automatic data logging.

⇒ Maximum safety even in high-pressure or temperature environments.

A Greener Approach: Microwave Assisted Reactions

In the chemical revolution of synthetic processes, green chemistry plays a significant role from the perspective of environmental safety (Chatel and Varma 2019; Varma 2014; C. J. Clarke et al. 2018; Tanaka and Toda 2000; P. T. Anastas, J. C. Warne 1998). To replace the traditional processes crucial in pharmaceuticals and fine chemicals, chemists need to create ecologically safe and sustainable alternatives (Kitanosono et al. 2018; Sarkar et al. 2016;

Hooshmand et al. 2019; Bruckmann, Krebs, and Bolm 2008; Varma 2016). Carbon-carbon (C-C) and carbon-heteroatom (C-X) bonds are formed and broken during organic synthesis (Salih 2022).

Therefore, creating novel techniques that address the demand for sustainable and environmentally friendly alternatives to fundamental processes is crucial for synthetic green chemistry research (Horváth 2018; Gawande et al. 2013; Li, Chao-Jun 2012; Ranu et al. 2012). Microwave (MW) radiation in synthetic organic chemistry has increased in popularity over the past several years, coupled with creating ecologically friendly techniques for sustainable chemistry; (Kokel, Schäfer, and Török 2017; Rathi et al. 2015; Kappe and Dallinger 2012; Mehta and Van der Eycken 2011; Kappe, C. Oliver, Doris Dallinger 2009).

Microwave synthesis was initially almost exclusively referred to as "MWAOS" (microwave-assisted organic synthesis). As this modern methodology gained popularity over time, additional chemistry disciplines were drawn to it, and numerous chemical industry businesses joined the community to benefit from the practical heating process, particularly in the method development and optimization process. However, the majority of the transformations carried out can still be regarded as organic chemistry, except for specific approaches in material sciences where metal oxides and zeolite-like materials are involved.

The presented examples comprise various reactions which are frequently performed under microwave irradiation such as:

- Metal-catalyzed reactions.
- Condensation reactions.
- Addition reactions.
- Substitution Reactions.
- Rearrangements.
- Synthesis of ionic liquids.
- Synthesis of nanomaterials.
- Synthesis of Polymer.

In organic synthesis, many examples of reactions have been described. Several studies on the use of microwave organic synthesis in solvent-free reactions, solid phase, phase transfer, non-classical solvent, multicomponent reaction, nanomaterials, polymers, heterocyclic chemistry, homogeneous and heterogeneous catalysis, medicinal and combinatorial chemistry, and green chemistry have been reported (Kappe et al. 2012).

We intend to provide an overview of some of the most recent trends in microwave-assisted organic synthesis, such as Metal-catalyzed reactions, condensation reactions, addition reactions, substitution reactions, and rearrangements.

Metal-Catalyzed Reactions

Metal-catalyzed reactions have been among the most well-known instances of MWAOS from the outset since the reaction rates of such transformations might be dramatically boosted. Every well-known cross-coupling and bond-forming reaction, including Heck, Suzuki, Negishi, Kumada, Hiyama, Sonogashira, and Ulmann (Figure 8) (Rahman et al. 2022) is well-suited to microwave irradiation and can be further accelerated by the use of metal catalysts. Additionally, the complexity of the reaction mixtures (e.g., reduced excess of chemicals, fewer mediators needed, etc.), as well as the amount of catalyst used could be decreased, when carried out under microwave irradiation. All of these reactions can be carried out using certain general procedures (shown in the schemes).

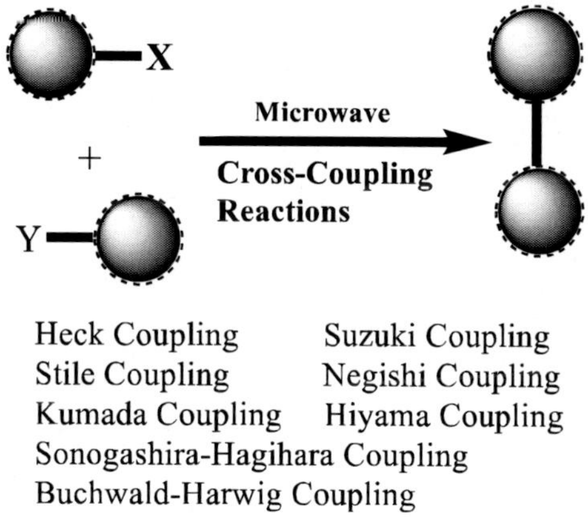

Figure 8. Various metal-catalysed reactions (Rahman et al. 2022).

Heck Coupling Reaction

The Heck-Mizoroki reaction is a mechanism for creating C-C bonds. Heck and Mizoroki first reported on it in the early 1970s, (Jagtap 2017; Knowles and Whiting 2007; Heck and Nolley 1972). Using a catalyst (transition metal, i.e., palladium) and a base (K_2CO_3, triethylamine), it carried out the coupling of alkyl, aryl, or vinyl halides with unsaturated compounds (such as alkenes, acrylates, etc.) under various conditions.

Reported (Rahman et al. 2022; Patel et al. 2019) (Scheme 1) in 2019, employing Pd NPs generated from rice husk-derived activated carbon-supported materials (nano-Pd@RAGC), a microwave-assisted Heck coupling reaction between an aryl halide and alkene. The most effective solvent for the reaction was polar ethanol. Within just 5 minutes, the required bi-arylated compounds were produced with up to 96% yield. The maximum yield, however, was only possible with two equivalents of K_2CO_3, which also served as a base. The reaction resulted in responses from aryl iodide and bromide. In addition, several substituted alkenes behaved smoothly under ideal reaction circumstances and produced the necessary products with good to outstanding yields. The used catalyst could be separated using only by filtration and reused up to 10 times without suffering any discernible catalytic activity loss.

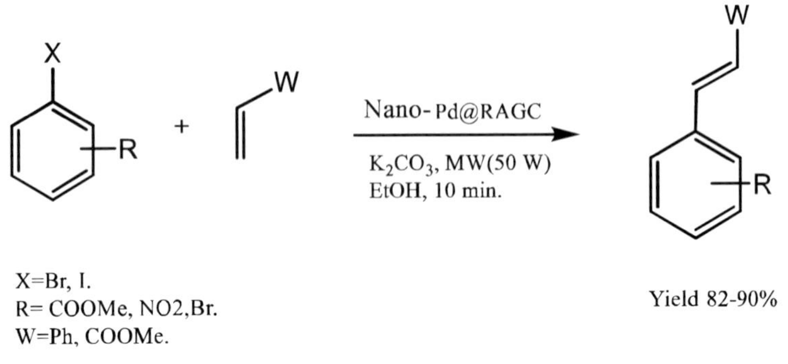

Scheme 1. Heterogeneous Pd-catalyzed microwave assisted Heck coupling reaction.

Pietro et al. (W Schruder, Christopher, Michael G Organ 2016) (Scheme 2) reported a microwave-initiated Heck coupling with Pd-coated silica macrospheres as a heterogeneous catalyst in batch processes as well as in continuous-flow organic synthesis. The reaction between aryl iodides and α,β-unsaturated esters produced the necessary coupling products in good yields in both setups.

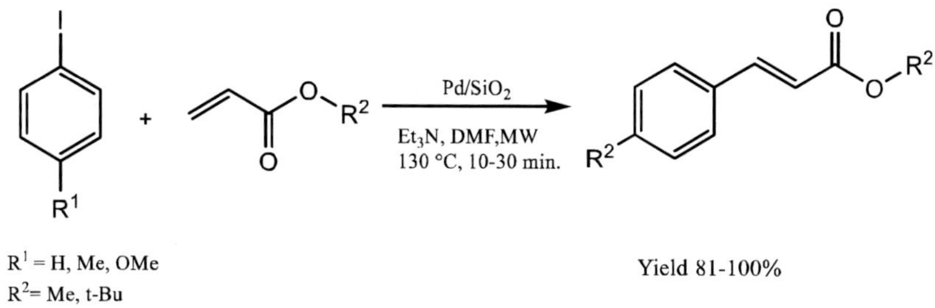

Scheme 2. Microwave-assisted heterogeneous Heck coupling reactions.

In 2012, Cravotto et al.(Cravotto et al. 2012) (Scheme 3) showed that Pd(II) immobilized on β-cyclodextrin, β-CD/Pd(II) acts as a catalyst in a microwave-promoted Heck coupling of styrene with aryl bromides. With a strong yield and excellent E/Z selectivity, these reactions produce selectively E-alkene. This technique delivered the corresponding required coupling products in 1 hour.

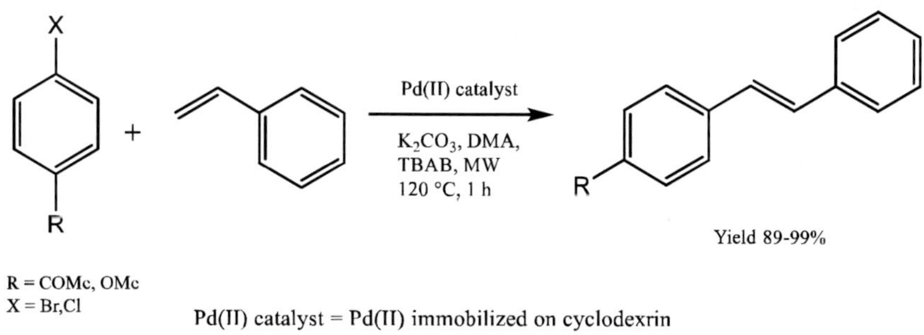

Scheme 3. MW assisted Heck reaction using β-CD/Pd (II) catalyst.

In 2011, Singh et al. (Allam and Singh 2011) (Scheme 4) demonstrated a phosphine-free, mild, and inexpensive MW-initiated approach for Heck reactions with great yields with NaOAc and TBAB in the presence of Pd(L-proline)$_2$ complex as catalyst.

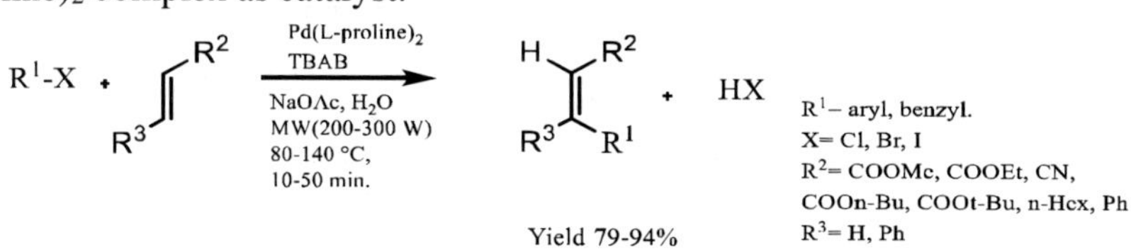

Scheme 4. Heck reaction of aryl halides with alkenes in aqueous medium.

Suzuki Coupling Reaction

One of the most effective synthetic methods in microwave-assisted organic synthesis, the Suzuki reaction, is a very effective method for carbon-carbon bond-forming reactions that produce biaryls through the cross-coupling procedure between aryl halides and aryl boronic acids (Martina et al. 2017; Larhed, M., C. Moberg 2002; Suzuki 1999; Larhed, Hallberg, and React 1996; Larhed 1996; Miyaura and Suzuki 1995; Chem1979).

Using a palladium catalyst in an aqueous media, Chakravarthy et al. 2019 (Rahman et al. 2022; Jeevan Chakravarthy et al. 2019 (Scheme 5) demonstrated a moderate and environmentally friendly Suzuki-Miyaura cross-coupling reaction between 1-chloro-2-tosylcyclohexene and aryl boronic acids to produce 1-chloro-2-arylcyclohexenes within 7 minutes, our technology delivered the desired goods at high yields. A number of substituted aryl boronic acids were tolerable and provided the necessary products with high regioselectivity and acceptable yields. The crystal structures of example products were used to corroborate the regioselective product constructions.

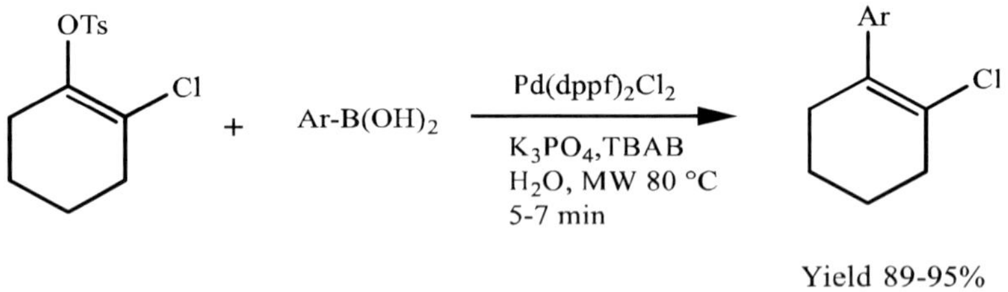

Scheme 5. Pd catalyzed C-C bond forming Suzuki coupling reaction.

By using a ligand-free Suzuki-Miyaura cross-coupling reaction between 5-iodo-2'-deoxyuridine and a variety of boronic acids in the presence of palladium and microwave irradiation, Len et al. (Gallagher-Duval et al. 2013) (Scheme 6) demonstrated a viable method for the synthesis of 5-aryl-2'-deoxyuridines. The reaction required a catalyst concentration of only 0.1 mol%. The highest yields of the intended products were produced when KOH was used as a base in the reaction. This procedure also made it possible to synthesize aromatic heterocyclic derivatives and compounds that were sterically hindered with good yields.

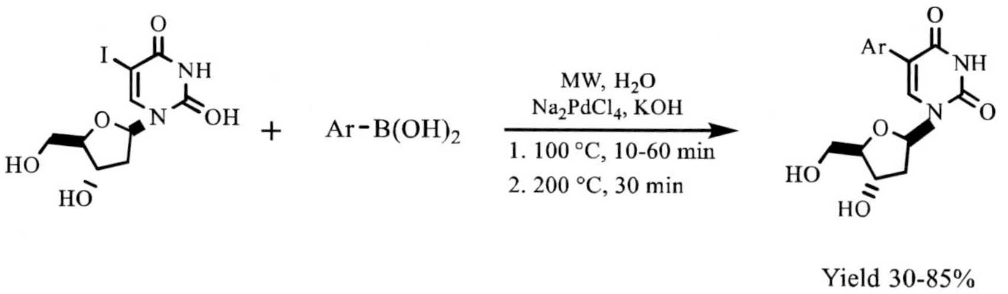

Scheme 6. Pd catalyzed C-C bond forming suzuki coupling reaction.

A Suzuki-Miyaura cross-coupling reaction of aryl boronic acids with different aryl sulfonates and halides was just recently Blanc group (Gallagher-Duval et al. 2013) (Scheme 7) reported, utilizing butanol and 5 mol% $PdCl_2(XPhos)_2$ as the catalyst. To produce the highest yield, however, one equiv. Tetrabutylammonium hydroxide was required as a base source. Within just 35 minutes, the requisite coupled products were produced with a 99% yield. Under optimal reaction conditions, a string of substituted phenyl tosylates phenyl boronic acids was engaged in the process. Furthermore, under the same reaction circumstances, heteroaryl boronic acids such as pyridin-4-ylboronic acid and thiophen-3-ylboronic acid provided the required compounds. In this reaction, even phenyl vinyl boronic acid was used, and up to 85% of the required compounds were produced. Alkyl boronic acid, however, did not participate in the reaction; instead, only degradation took place.

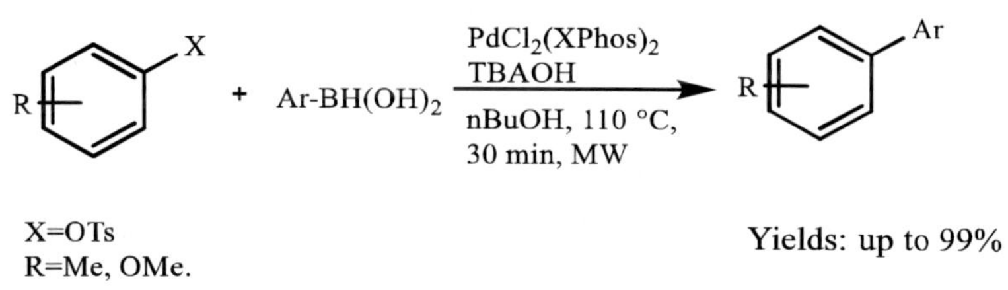

Scheme 7. Pd catalyzed C-C bond forming Suzuki coupling reaction.

Stille Cross-Coupling Reaction

The incorporation of an organotin moiety together with palladium catalysts is the distinguishing characteristic of the Stille cross-coupling reaction (also known as the Migita-Kosugi-Stille coupling) (Cordovilla et al. 2015; Gallagher-Duval et al. 2013).

Makhloutah et al. reported Stille reactions involving (hetero) arylstannanes and nitro-PDI derivatives (Makhloutah et al. 2022). The use of nitro-PDI compounds in coupling reactions facilitated by palladium. We started our research using the standard Stille cross-coupling of tributyl phenyl stannane and nitro-PDI. Addition of LiCl and TFP additives to standard Stille cross-coupling under microwaves at 150°C for 2 h resulted in the reaction yield production of up to 98% (Scheme 8).

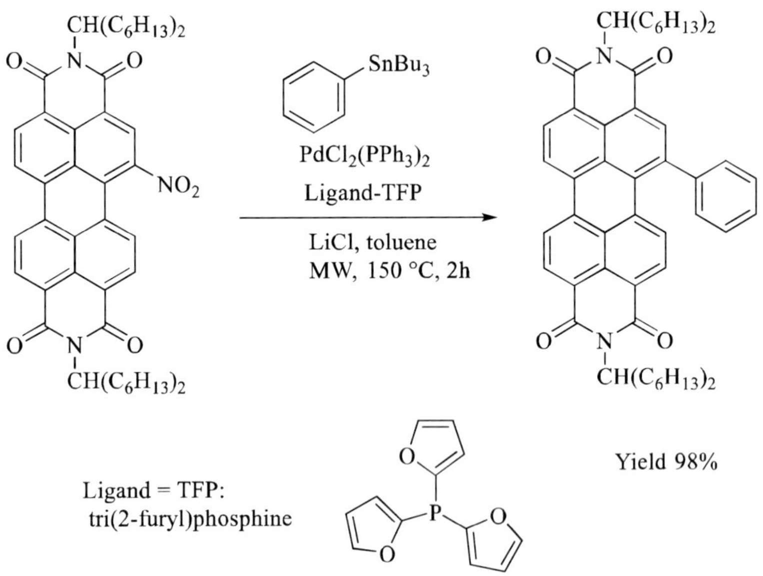

Scheme 8. Stille-type reaction in perylenediimide(PDI).

Salih and Baqi has demonstrated the use of palladium nanoparticles made *in situ* from $PdCl_2$-PEG400 in the presence of DMAP to synthesize many functionalized biaryls (Salih and Baqi 2019; Xu et al. 2015). Under microwave heating, one equivalent of tetraphenyl stannane was combined with aryl halides to produce the corresponding biaryls in high yields (Scheme 9). Biphenyl was occasionally formed as the single byproduct from the homocoupling of tetraphenyl tin. In general, aryl halides with electron-withdrawing substituents provided better yields than aryl halides with electron-donating substituents.

The interaction of $SnPh_4$ with 1-bromo-4-chlorobenzene proved that the catalyst could be used again. The residual PEG-400 and Pd nanoparticles were put through another run after the organic material had been extracted using diethyl ether. Even though the catalyst was reused four times and the reaction time extended to 40 minutes, the yield dropped from 94% to 81%.

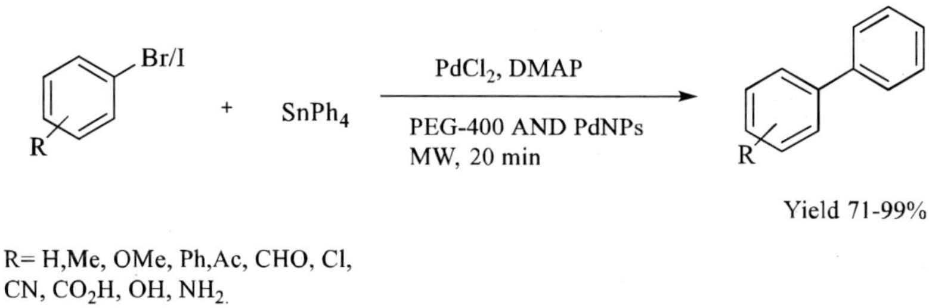

Scheme 9. In situ generated Pd nanoparticles mediated Stille coupling reactions.

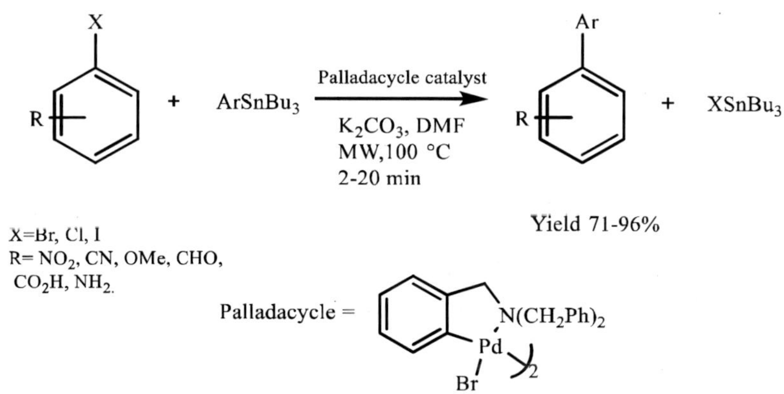

Scheme 10. Pd catalyzed C-C bond forming stile coupling reaction.

Hajipour group (Hajipour, Karami, and Rafiee 2012) developed a straightforward and practical MW-promoted Stille cross-coupling reaction involving various aryl halides and phenyl tributyltin in the presence of 0.3 mol% of $[PdC_6H_4(CH_2N(CH_2Ph)_2)(m\text{-}Br)]_2$ complexes in DMF at 100°C (Scheme 10). The employed base K_2CO_3 served as a co-catalyst, accelerating the reaction by reducing the palladium(II) complex. By weakly covalently bonding with the palladium moiety and functioning as a microwave-active polar solvent, DMF produced the required compounds with high yields in a short amount of time. The iodo- and bromo substituted were more reactive in comparison to chloro-substituted aryl derivatives. In a final instance, the Stille coupling reactions produced the desired products (substituted biaryls) in somewhat longer periods and with modest yields.

The Stille reaction was one of the first transition metal-catalyzed reactions to be accelerated with microwave heating in single-mode irradiation (Larhed 1996) (Scheme 11).

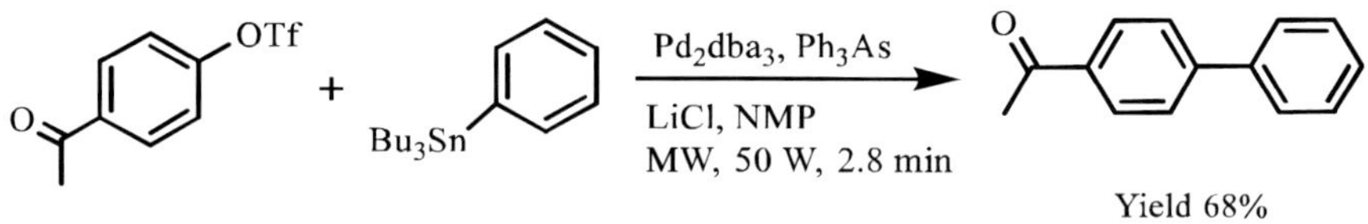

Scheme 11. Stille coupling in solution with 4-acetylphenyl triflate.

Negishi Coupling Reaction

The first successful microwave-assisted cross-couplings of organozinc compounds used both aryl- and alkyl zinc bromides in a shorter time. Despite the importance of this general C-C bond-forming reaction, reports on microwave-induced Negishi couplings are limited.

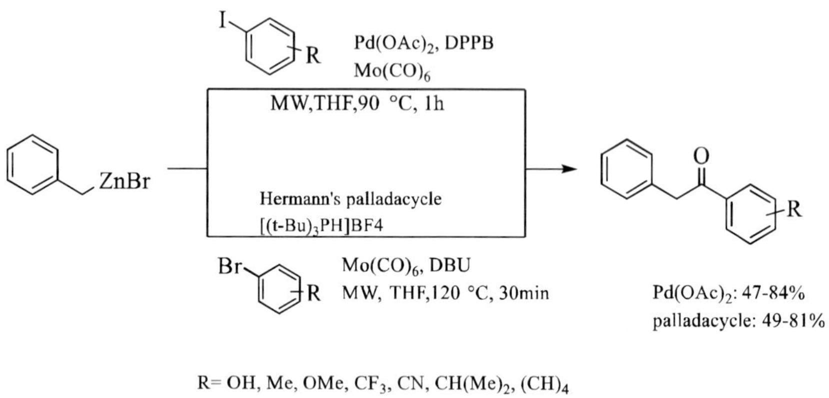

Scheme 12. Synthesis of biaryls *via* three-step Negishi cross-coupling reactions.

Motwani and Larhed, under controlled microwave irradiation, created carbonylative Negishi cross-couplings for aryl iodides and aryl bromides using benzylzinc bromide. Molybdenum hexacarbonyl is used as a solid *in situ* source of CO in the carbonylative cross-coupling, which involves adding CO between the coupling partners. Based on the aryl halide employed, the reaction conditions were tuned, as shown in Scheme 12. It produces many diarylated ethanones, a frequent pharmacophore, in moderate to high yields (Salih and Baqi 2019; Motwani and Larhed 2013).

The synthesis of sterically hindered chiral binaphthalene derivatives has a procedure that the Espinet group publicly disclosed in 2007. $Pd_2dba_3CHCl_3$ and a chiral ligand, (R,Sp)-(2-diphenylphosphino-ferrocenyl)ethyl-dimethylamine, were used to conduct the enantioselective cross-coupling reactions between naphthylbromides and bis(naphthalen-1-yl)zinc compounds (Scheme 13). The coupling was successful even with the sterically challenging substrate in THF at 100°C for 45 min in an inert environment ($R_1 = R_2 =$ Me) (Salih and Baqi 2019; Genov, Almorín, and Espinet 2007). In addition, identical syntheses under Suzuki reaction conditions (using 1-naphthaleneboronic acids rather than bis(naphthalen-1-yl)zinc derivatives) was carried out provides slightly better results than the Negishi technique.

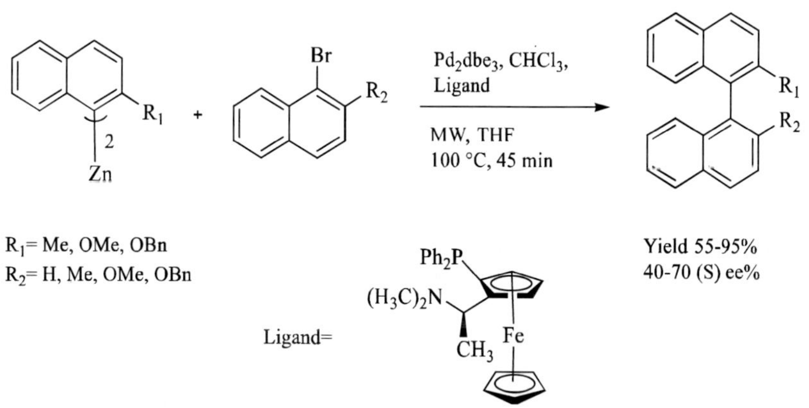

Scheme 13. Synthesis of chiral binaphthalene derivatives *via* Negishi cross-coupling reaction.

Using aryl chlorides or aryl bromides with magnesium turnings while being microwave-irradiated is an effective method for producing aryl zinc compounds for the Negishi reaction. Following a transmetalation using $ZnCl_2$-TMEDA, the aryl zinc species was created, and it was then immediately added to a coupling reaction with aryl bromide derivatives utilizing $Pd(PPh_3)_2Cl_2$ as a precatalyst. Depending on the substrates employed, the cross-coupling reaction was finished quickly after exposure to MW at 80° or 120°C in THF or a THF-DMF mixture. For example, under the conditions mentioned above, aryl chlorides did not react; however, at normal temperatures, the coupling reaction between 4-bromophenylzinc chloride and bromobenzaldehyde took place. Utilizing 2-chloro-bromobenzene, where the cross-coupling occurred selectively via bromide and provided the matching biaryl, the chemo-selectivity of the reaction was evident. (Salih and Baqi 2019; Mutule and Suna 2005) (Scheme 14).

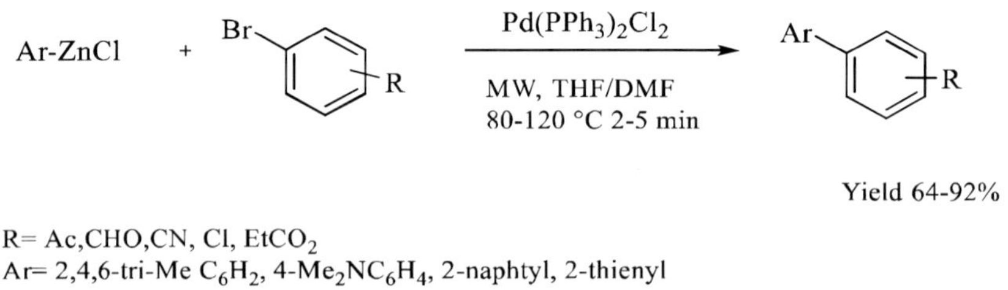

Scheme 14. Synthesis of biaryls *via* Negishi coupling reaction.

The aryl bromides or iodides were combined with activated Rieke zinc to create the organozinc reagents. Electron-rich aryl chlorides got deactivated, could not drive nickel-catalyzed reactions to completion, and resulted in large levels of homocouplings. However, palladium was found to be more efficient with both electron-rich and electron-poor aryl chlorides when combined with electron-rich phosphines (Walla and Kappe 2004) (Scheme-15). Additionally, the products of coupling n-butylzinc chloride and resin-bound aryl chlorides could be separated effectively in high yields.

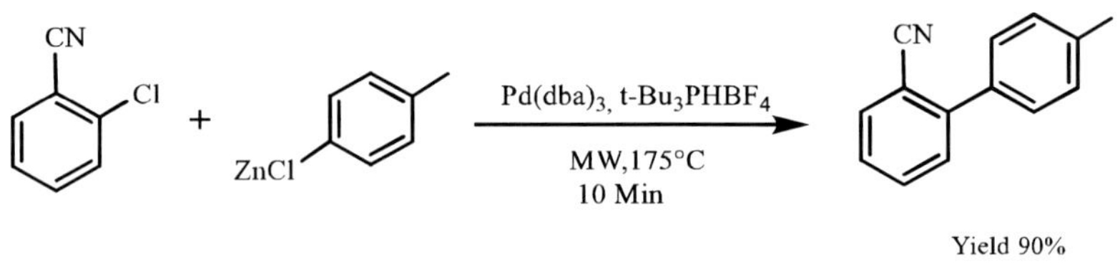

Scheme 15. Pd catalyzed C-C bond forming Negishi coupling of an aryl chloride reaction.

A single-mode microwave reactor was selected for the scale-up work. The optimized reaction conditions involved heating to 160°C for 1 min in THF in the presence of $PdCl_2(PPh_3)_2$ as a catalyst providing an isolated yield of 90% of the desired biaryl (Stadler et al. 2003) (Scheme 16).

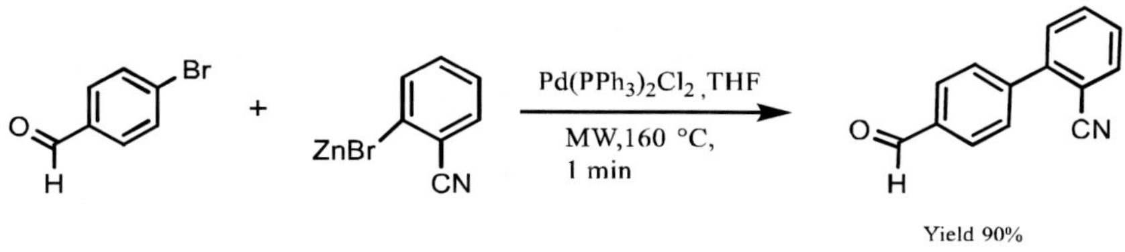

Scheme 16. Pd catalyzed C-C bond forming Negishi coupling of an aryl bromide reactions.

The large-scale synthesis was carried out in two vessels under an argon atmosphere in the multimode batch reactor. The reaction was heated to 160°C, employing a heating ramp of 2 min, and kept at 160°C for an additional minute, otherwise using identical reaction conditions as in the reported small-scale experiment.

Purification of the combined reaction mixtures by flash chromatography provided biaryl in a 77% isolated yield.

Kumada Coupling Reaction

Microwaves were utilized both in the preparation of the Grignard reagent and in Kumada coupling (Walla and Kappe 2004; Hassan et al. 2002) with aryl chlorides. But, homocoupled side products were observed in microwave as the heating source.

Due to the sensitivity of the organomagnesium halides (Grignard reagents) as coupling partners, the Kumada reaction is one of those methods that is rarely documented in the peer-reported literature, especially under microwave circumstances. However, the synthesized oxime-based palladacycle precatalyst was used in coupling reaction between various combinations of alkyl halides and Grignard reagents at extremely low catalyst loading, resulting in high yields of the predicted product (Salih and Baqi 2019) (Scheme 17). Under optimal reaction conditions, the recyclability of the catalyst was tested using bromobenzene and phenylmagnesium bromide. In the fourth run, a black precipitate formed due to Pd leaching and caused a yield drop from 91% to 63%.

R_1-MgBr + X-R_2 $\xrightarrow[\text{THF, MW 100 °C, 1-12 min}]{\text{Pd-Cat.}}$ R_1-R_2

Yield 76-98%

X= Br
R_1= vinyl, Ph
R_2= n-hexyl, Ph, 4-$MeOC_6H_4$, 4-$NO_2C_6H_4$,
5-pyrimidihyl, 9-anthracenyl, etc.

Scheme 17. Paladacycle-catalyzed Kumada coupling reactions.

The cross-coupling reaction of mono- and di-iodinated carba-closo-dodecaborate anions with various Grignard reagents have been studied by Finze's group. Using $PdCl_2P(Ph_3)_2$ in THF at roughly 60°C, the coupling products were produced under microwave conditions in high yields and with reduced reaction durations (Salih and Baqi 2019; Himmelspach, Reiss, and Finze 2012) (Scheme 18).

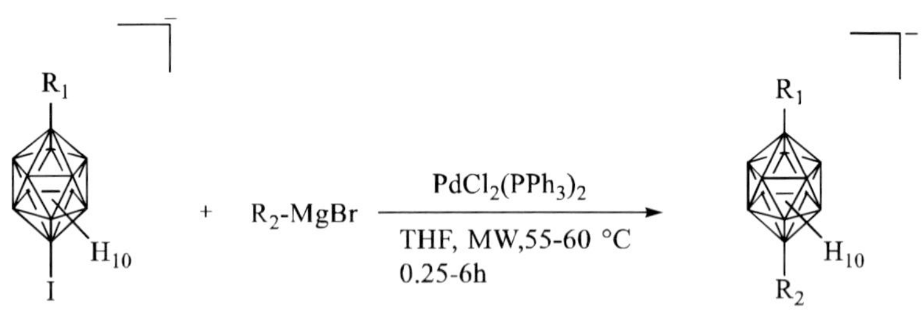

Yield 61-90%

R_1= H, Me, Ph
R_2= Me, CH_2SiMe_3, $CCSiEt_3$, CCPh, etc.

Scheme 18. Microwave-assisted Kumada-type cross-coupling reactions.

Grignard reagents have been efficiently, abundantly, and reproducibly produced from passive aryl chlorides and bromides. It produce the initial aryl magnesium halide and to promote a subsequent Kumada coupling in the synthesis of a novel HIV-1 protease inhibitor in microwave irradiation (Larhed and Nilsson 2005) (Scheme 19).

Kumada coupling of Grignard reagent with aryl chlorides. But, homocoupled side products are observed in microwave as the heating source (Walla and Kappe 2004) (Scheme 20).

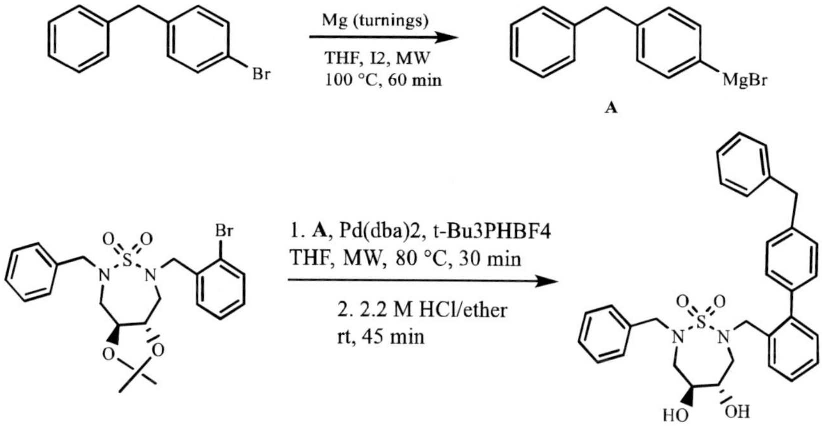

Yield 67%

Scheme 19. Preparation of a cyclic HIV-1 protease inhibitor by Grignard and Kumada chemistry.

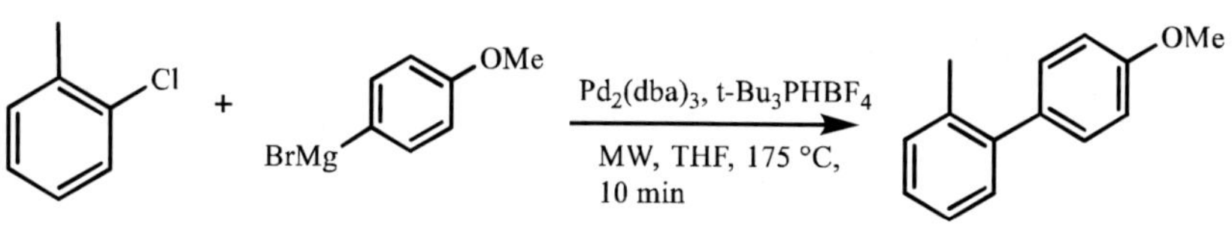

Yield 94%

Scheme 20. Kumada reaction with aryl chlorides.

Hiyama Coupling Reaction

The Hiyama coupling reaction, which uses palladium as a catalyst, produces carbon-carbon bonds (C-C) by reacting organosilanes with organic halides.

Clarke reported the first microwave-accelerated Hiyama coupling (M. L. Clarke 2005; Negishi, Ei-ichi, and Armin De Meijere 2003) (Scheme 21). It was noted that the organosilicon reactants are particularly desirable in synthesis due to their availability and benign qualities, but their potential is constrained by their poor nucleophilicity. An electron-rich N-methylpiper-azinecyclohexylphosphine ligand permitted aryl bromides and activated aryl chlorides to react under palladium catalysis under microwave irradiation.

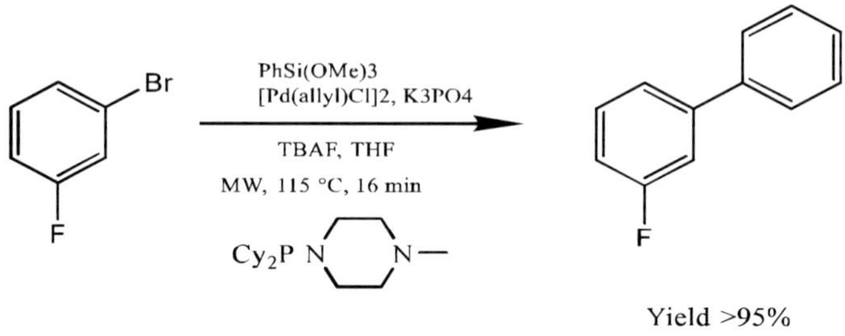

Scheme 21. Hiyama reaction with an aryl bromide.

1-Methyl-2-aminoalkylimidazole derivative was reported as an efficient ligand for the Hiyama reaction under solvent-free conditions in the presence of $Pd(OAc)_2$. Aryl bromides with a range of electron-donating and electron-withdrawing substituents were cross-coupled with aryl(trimethoxy)silane, affording the correspondent biaryls in good to high yields under MW irradiation (Scheme 22). The congested 2-bromotoluene yielded the coupling product at 85%, while the 2-bromophenol coupled with 40% yield. These were rationalized for the poor reactivity, as the 2-bromophenol could deactivate the catalyst by coordination with the metal center. The reaction with heteroaryl halides and aryl chlorides was generally less effective (Salih and Baqi 2019; Martínez, Pastor, and Yus 2014). Moreover, a variety of hydroxy- and amino-functionalized imidazoles were also evaluated as efficient ligands for the palladium-catalyzed Hiyama reaction.

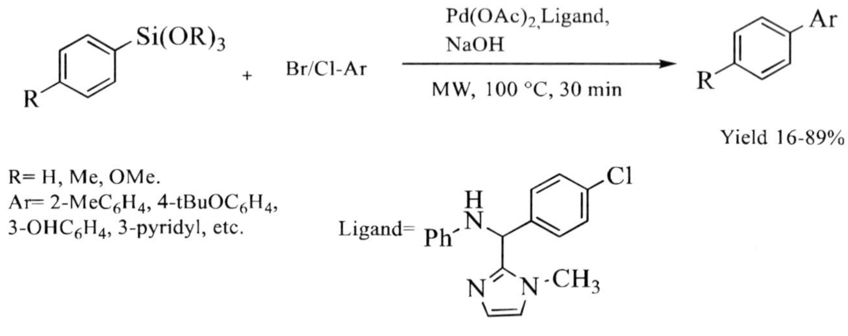

Scheme 22. Palladacycle-catalyzed Hiyama coupling reactions.

The supported PdNPs catalyzed sodium hydroxide activated Hiyama cross-coupling reaction of phenyltrimethoxysilane with various bromo and chloroarenes under MW irradiation was reported by Kaur et al. in 2012 (Shah and Kaur 2014) (Scheme 23). The reaction was carried out without any extra ligands, surfactants, or hazardous solvents. The catalyst made of Pd NPs could be utilized repeatedly.

Scheme 23. Hiyama cross-coupling of aryl halides with phenyltrimeth oxysilane.

Aryl halides and alkenyltrialkoxysilanes were reported to undergo microwave-assisted Hiyama reactions in 2008, with TBAB acting as the promoter in a ligandless environment (Scheme-24). Vinyltrimethoxysilane was discovered to be an efficient coupling partner for the formation of various styrene derivatives from palladacycles derived from 4-hydroxy acetophenone oxime (Alacid and Nájera 2008).

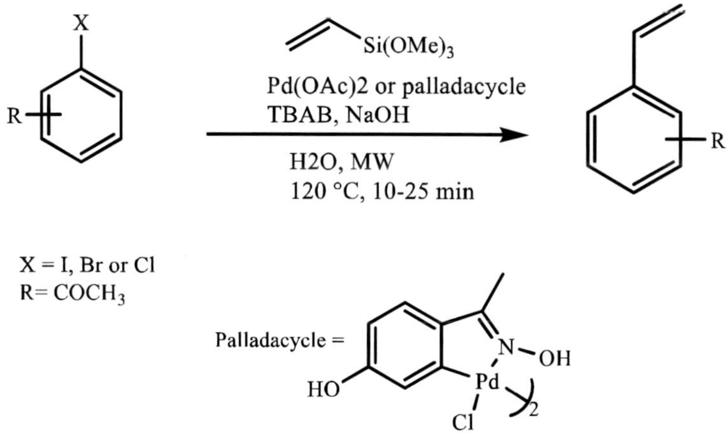

Scheme 24. Synthesis of Styrenes *via* Hiyama cross-coupling reactions.

Sonogashira–Hagihara Coupling Reaction

Terminal alkynes and aryl halides (Erdélyi, Máté, and Adolf Gogoll 2003; Sonogashira 1975) interact through the Sonogashira coupling process. Sonogashira cross-coupling process is often carried out in polar and high boiling point solvents, such as DMSO, DMF, or N-methyl-2-pyrrolidone (NMP), with a palladium catalyst, phosphine ligands, and copper/copper salts acting as co-catalysts (Doucet and Hierso 2007; Chinchilla and Na 2007; Elangovan, Wang, and Ho 2003; Hundertmark et al. 2000).

It was first assessed using the outcomes of various MW irradiation settings using the easily accessible synthesis partners 5-chloro-8-hydroxy-7-iodoquinoline and phenylacetylene as examples. Sodium aluminate, zinc powder (2%), chloroform as the solvent, and the halo functionality of 8-hydroxyquinolines were combined in an RB flask. The organic layer evaporated under less pressure. The mixture was agitated at RT for ten to fifteen minutes before adding an aromatic alkyne (Saha and Mukherjee 2022) (Scheme 25). The coupling products were generated under microwave conditions in high yields in a flask that had a partition and was microwaved for five minutes at 140°C (250 W).

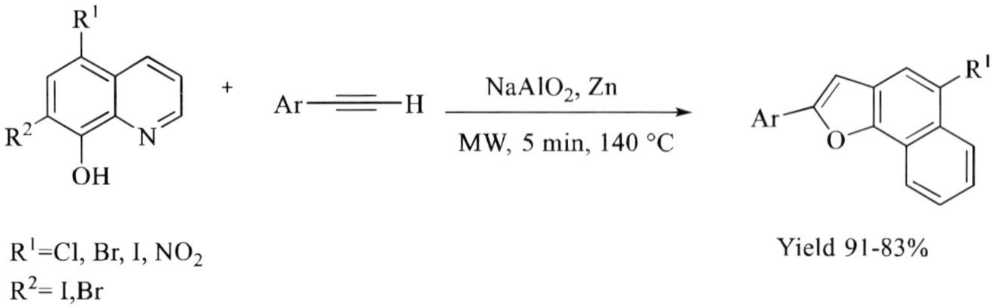

Scheme 25. Synthesis of furo[3,2-h] quinoline by sodium aluminate, zinc catalyzed under MWR.

Gedu and his coworkers (Karu and Gedu 2018) reported an intermolecular Sonogashira coupling reaction using Pd-catalyst followed by domino Heck cyclization to produce alkyne-substituted dihydrobenzfurans (Scheme 26). This process displayed a wide range of acetylene substrates, and all were given the matching products in excellent yields in a relatively shorter time. Under microwave irradiation, just 1mol% of $Pd_2(dba)_3$ was required to produce up to 98% of the necessary compounds in 15 minutes.

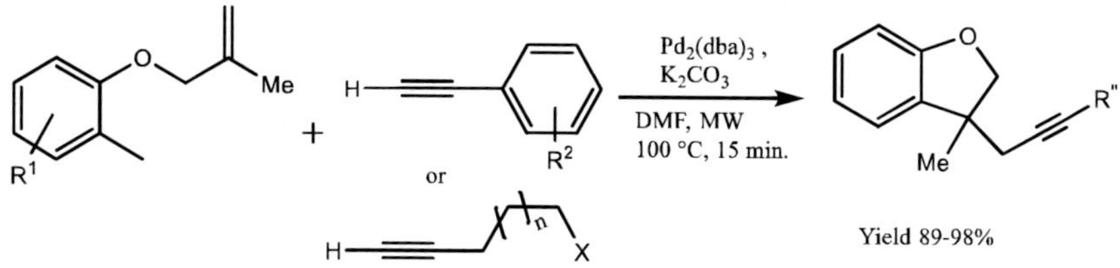

Scheme 26. Heck cy cyclization followed by intermolecular Sonogashira coupling reaction using Pd-catalyst.

Using SiliaCatPd(0)(Si-Pd-1) as the catalyst, CTAB as the surfactant, and water as the solvent, Najera et al.(Cívicos, Alonso, and Nájera 2013) described a copper-free microwave-mediated oxime Sonogashira coupling synthesis of various aryl imidazole 1-ylsulfonates with aryl-substituted terminal alkynes (Scheme 27). Lower yields were offered by other surfactants such as SDS, PTS, TBAOH, SPAN80, etc. Under the ideal reaction conditions, several substituted terminal alkynes and aryl imidazol-1-ylsulfonates provided the required products. More intriguingly, at optimal reaction conditions, the alkyl-modified terminal alkynes also reacted to the reaction and produced the expected coupled products with satisfactory yields.

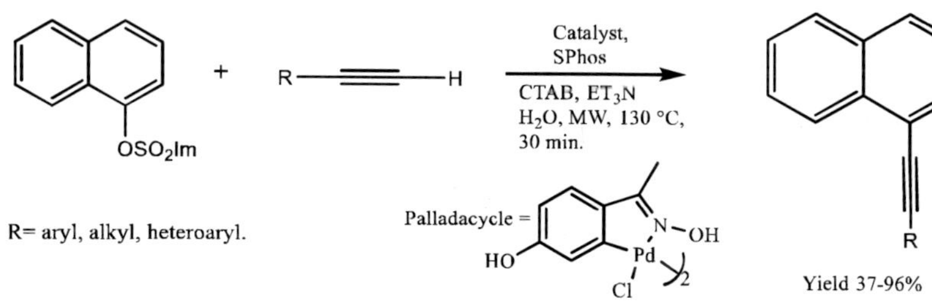

Scheme 27. Microwave-mediated palladacycle catalysed Sonogashira-Hagihara couplings of various alkynes in water.

The copper-catalyzed Sonogashira coupling reaction of aryl iodides with terminal acetylenes in a water-polyethylene glycol solvent system was demonstrated by Chen et al. in 2010 (Chen et al. 2010) (Scheme 28) and was shown to be a very helpful approach. With good to outstanding yields, the required products were produced by alkylating a range of aryl iodides with both electron-rich and -deficient substituents under MW heating or by heating in an oil bath. With little loss in yields, the catalyst was re-used numerous times.

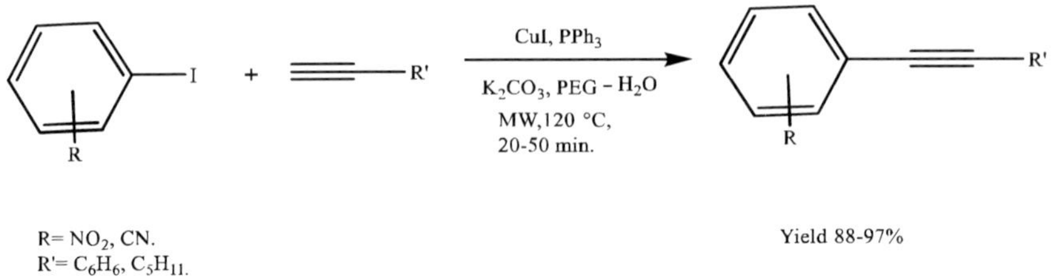

Scheme 28. Copper-catalyzed Sonogashira coupling reaction of aryl iodides.

Buchwald-Hartwig Coupling Reaction

The palladium-catalyzed Buchwald-Hartwig reaction (Scheme 29) of aryl bromides and aniline has been widely used for the formation of C-N bond that led to substituted anilines.

For the optimization, o-bromo-benzoic acid and aniline were chosen as a model substrate for Buchwald-Hartwig reaction and performed the reaction under microwave irradiation, provided product with 90% yield. Under microwave irradiation such as tBuOH and KOt-Bu was found to be the most suitable base, were more efficient and produced the maximum amount of product in 10 minutes. (Khan, Parveen, and Sharma 2021) (Scheme 29).

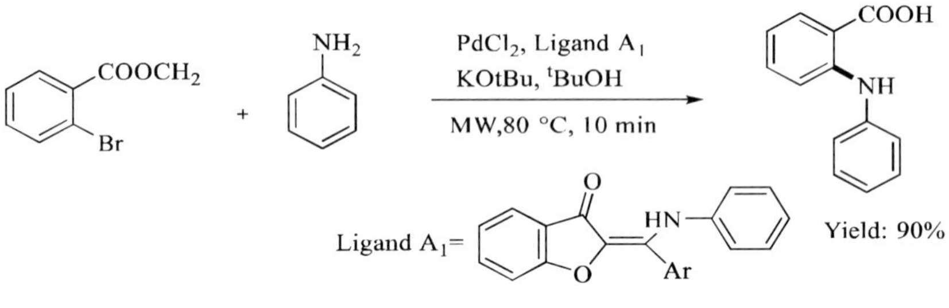

Scheme 29. Buchwald-Hartwig coupling reactions.

In addition to the normal cross-coupling reactions that establish carbon-carbon bonds, palladium catalysis can also be used to form carbon-nitrogen bonds to produce important heterocyclic scaffolds. The following (Scheme 30) represents a solvent-free method for producing pyridopyrimidines in a single pot using Buchwald-Hartwig amination and cycloaddition reaction (Saikia et al. 2015).

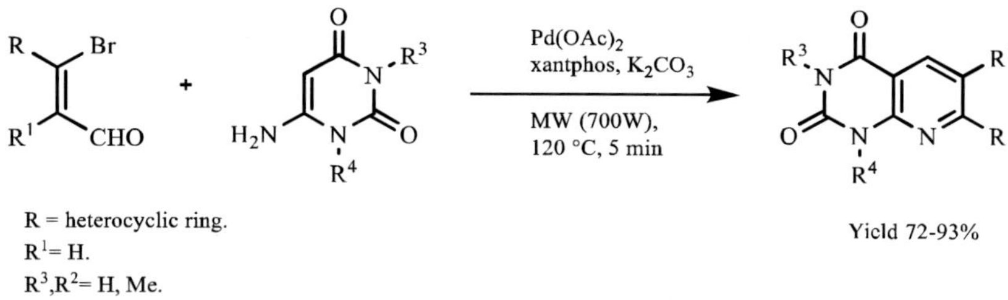

Scheme 30. Synthesis of pyrido [2,3-d] pyrimidine library via Buchwald-hartwig amination.

Carbon-Heteroatom Bond Forming Reactions

A fascinating and significant area of microwave-assisted organic chemistry is the creation of carbon-heteroatom bonds (Lai, Ghosh, and Hajra 2021; Escudie, Ranaivonjatovo, and Rigon 2000). Over the decades, the creation of carbon-heteroatom bonds has been greatly influenced by transition metal-catalyzed processes (Girard et al. 2012; Beccalli et al. 2007).

Carbon-Nitrogen Bond-Forming Reactions

The formation of C-N bonds is an essential and powerful technique for synthesizing biologically and medicinally important hetero compounds (Evano, Blanchard, and Toumi 2008).

Jiang et al. (Jiang and Huang 2013) described a very useful microwave-promoted method for synthesizing aryl aminomethyl acetylenes by the reaction of aromatic boronic acids with ammonia and propargyl halides (bromide or chloride) in a water medium using copper(I) oxide as a catalyst (Scheme 31). Aromatic boronic acids with electron-releasing groups reacted more efficiently than those with electron-withdrawing groups. Due to steric effects, the substrates with ortho-substituents produced the corresponding products with lower yields.

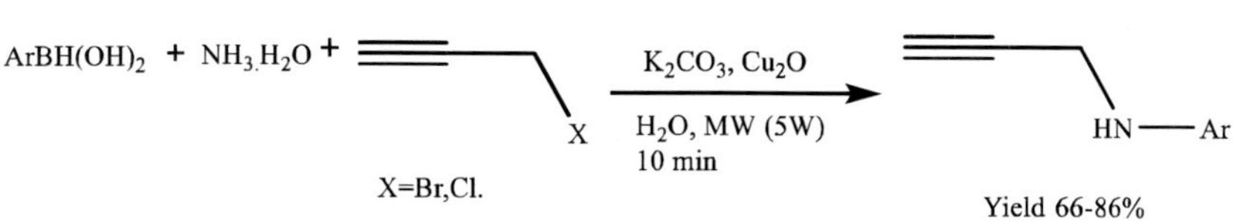

Scheme 31. The synthesis of arylaminomethyl acetylenes C-N bond form in under microwave irradiation.

Chilin et al. (Marzaro, Guiotto, and Chilin 2009) demonstrated a greener, catalyst-free method for producing mono-N-alkylated products from the reaction of aromatic amines with alkyl halides in a water medium using microwave irradiation (Scheme 32). This synthetic method worked well with a wide range of arylamines and also with deactivated amines; it produced only mono-N-alkylated products and was chemoselective in the presence of multiple functional groups. Under standard reaction conditions, alkyl iodides, bromides, and even chlorides were successfully involved in the reaction. However, yields were lower in the case of alkyl chloride when compared to alkyl iodides.

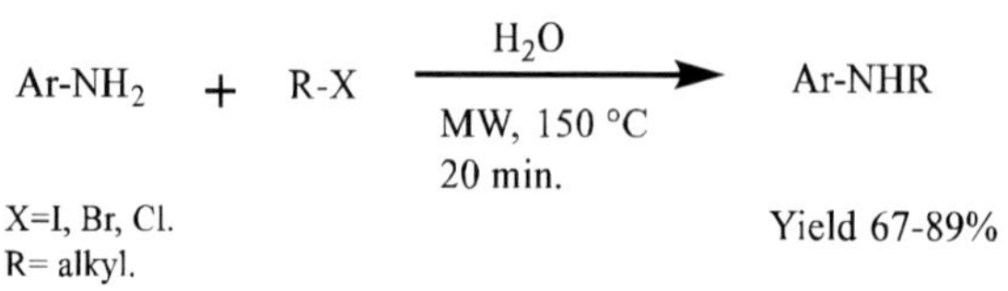

Scheme 32. Synthesis of mono-N-alkylated products under MW irradiation.

Carbon-Oxygen Bond-Forming Reactions

Diaryl ether-containing compounds are very important because they are found in a variety of biologically and medicinally active compounds (Boger et al. 1999; Toda et al. 1998).

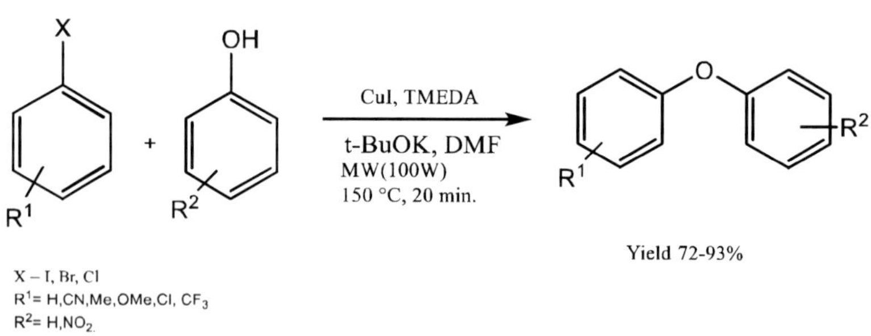

Scheme 33. C-O bond formation in Microwave-initiated Ullmann-type coupling reactions.

Zhu et al. reported a mild and general microwave-initiated protocol for the Ullmann C-O bond formation reaction of a broad array of activated and unactivated aryl bromides and iodides with phenols using cesium carbonate or t-BuOK as the base in anhydrous DMF or NMP (Zhu et al. 2007) (Scheme 33). All halides, electron-poor, electron-neutral, and electron-rich aryl halides produced high-yielding diaryl ethers.

Carbon-Phosphorus Bond-Forming Reactions

Due to their bioactive qualities, organophosphorus compounds are crucial in the fields of industrial and medical chemistry (Engel 1977). Carbon-Phosphorus (C-P) bond-forming processes are used to create the organophosphorus chemicals. (Jablonkai and Keglevich 2015; Taylor et al. 2015).

In the year 2012, Rummelt et al. (Rummelt, Ranocchiari, and Bokhoven 2012) (Scheme 34) reported a relatively benign and efficient MW-initiated method for cross-coupling reaction of various iodo- and bromobenzoic acids with diphenylphosphine oxide in a water medium without any ligand or other additives using heterogeneous and recyclable Pd/C catalyst. Under MW irradiation, the various halogenated benzoic acids produced the desired products, whereas conventional heating did not.

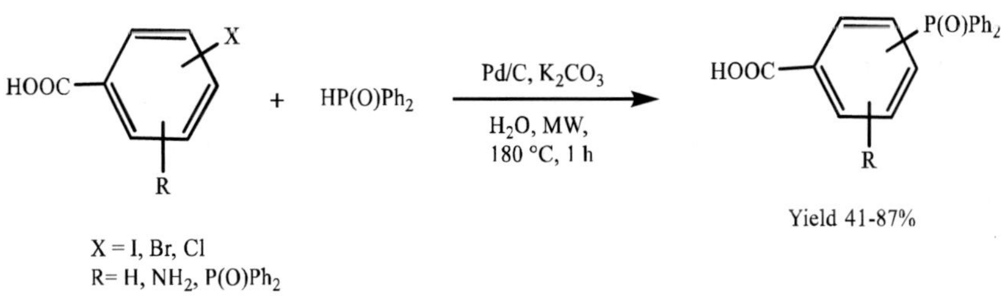

Scheme 34. C-P coupling of several halogenated benzoic acids under MW heating.

Carbon-Sulphur Bond-Forming Reaction

In microwave-assisted organic synthesis, the cross-coupling reaction between aryl halides and sulfur nucleophiles is very efficient in generating transition metal-catalyzed C-S bonds.

The synthesis of aryl sulfides from the substitution reaction of aromatic thiols with aryl halides using water-TBAB as a solvent was demonstrated by

Velmathi et al. (Vijayaraghavan, Pal, and Vinu 2010) (Scheme 35) in the year 2010. The presence of bases like Na_2CO_3, K_2CO_3, and NaF made the reaction possible. NaF as the base, resulting in the highest yields. Further, the TBAB-H_2O mixed solvent produced higher products than any individual solvent. Furthermore, raising the temperature from 150°C to 170°C had no discernible impact on improving reaction yields. Compared to traditional heating techniques, this method produced the necessary products with a high yield and shorter time.

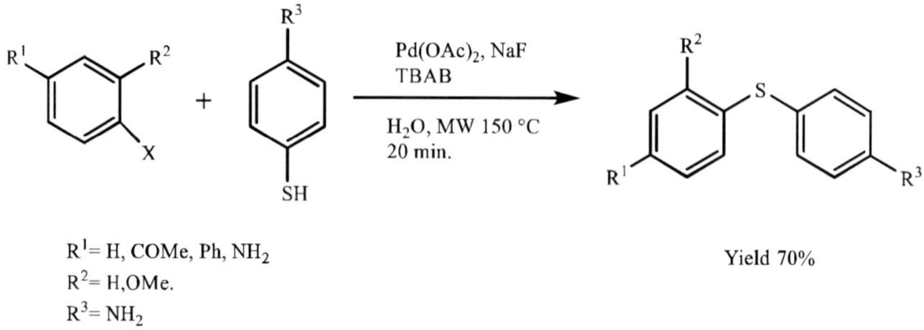

Scheme 35. C-S bond form in MW-initiated coupling reaction of aryl halides with aryl thiol.

Condensation Reactions

A wide variety of MW-assisted aldol (Arnáiz and Aguado 1995; AbenhaïM et al. 1994) and Knoevenagel condensation reactions have been accomplished using relatively benign reagents such as ammonium acetate (Mitra, De, and Karchaudhuri 1999), including the (Pii et al. 1996) Condensation reactions are frequently employed to produce carbon-carbon and carbon-heteroatom bonds in organic and medicinal chemistry. However, in some circumstances, when reactions are carried out in sealed vessels, it is necessary to take into account the generation of volatile by-products in order to maximize conversion. A proper dilution and filling volume must be calculated before the reaction to prevent exceeding the operating limits of the used reaction containers. Numerous condensation processes have been carried out successfully in each of the several microwave reactors using this knowledge.

Using substantially less power (70 W) than conventional heating techniques, microwave heating produces yields comparable to current practice in about 5 minutes. With N, N′-(2,4,6- trimethylphenyl)imidazolium chloride

(IMesHCl) as the catalyst and 1,8-diazabicyclo[5.4.0]undec-7-ene (DBU) as the base, the best yields were obtained (Morgan et al. 2022) (Scheme 36).

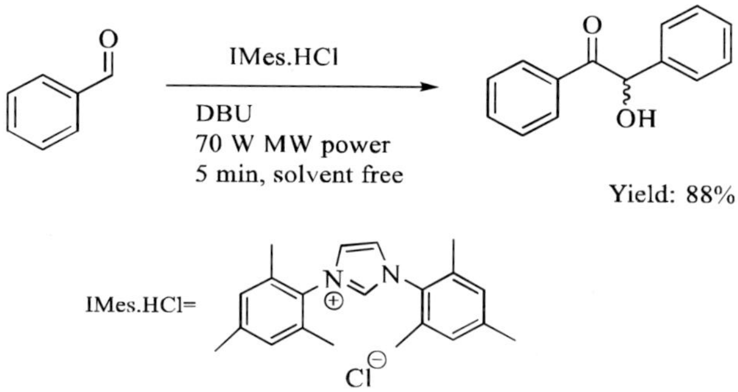

Scheme 36. Benzoin condensation of benzaldehyde under microwave irradiation.

The Biginelli cyclocondensation, which is straightforward one-pot multicomponent reaction that provides access to a wide range of useful dihydropyrimidine scaffolds. Under relatively benign conditions, the reaction may tolerate a wide range of aldehydes, different -keto compounds, and a several urea/thiourea derivatives to produce the necessary heterocycles (Scheme 37) (A. Stadler, C. O. Kappe 2001). As a result, the reaction is frequently employed to generate parallel libraries since a single temperature/time profile yields acceptable results for a large number of building block combinations. The dihydropyrimidine core can be altered by a many transformations in order to further increase the number of derivatives. The reaction can be carried out at reasonably high concentrations and generate exceptional yields of more than 100 grams each run when scaled up in parallel or batches.

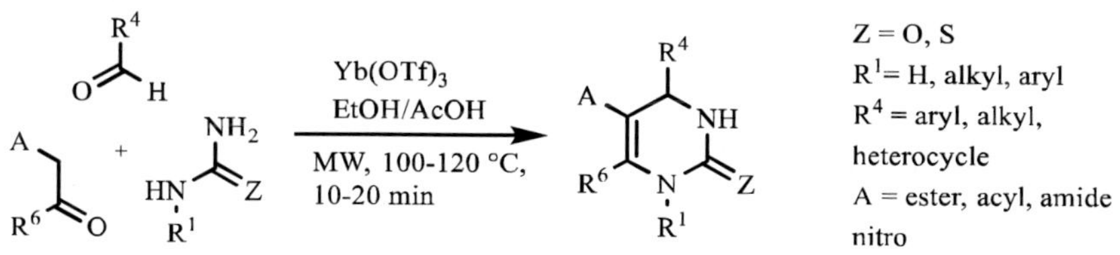

Scheme 37. Bigenelli cyclocondensation.

The Knoevenagel condensation of aldehydes and ketones is a different kind of condensation reaction and thus a subset of an aldol reaction (Treu et al. 2008) (Scheme 38). Through the Knoevenagel condensation, significant intermediates like oxindoles can also have their scaffolds decorated.

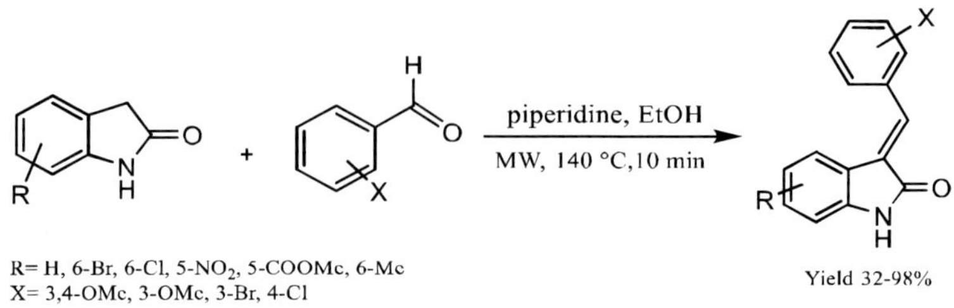

Scheme 38. Decoration of oxindole scaffolds by Knoevenagel condensation.

Thioamides are also a crucial building element in pharmaceutical research. These structures are easily accessible and allow for Kindler condensation using elemental sulphur (A. Stadler et al. 2003) (Scheme 39-a). While sealed vessel microwave synthesis enables the rapid synthesis of thioamide building blocks in a very short time, conventional Kindler condensations require substantial reaction durations and high temperatures.

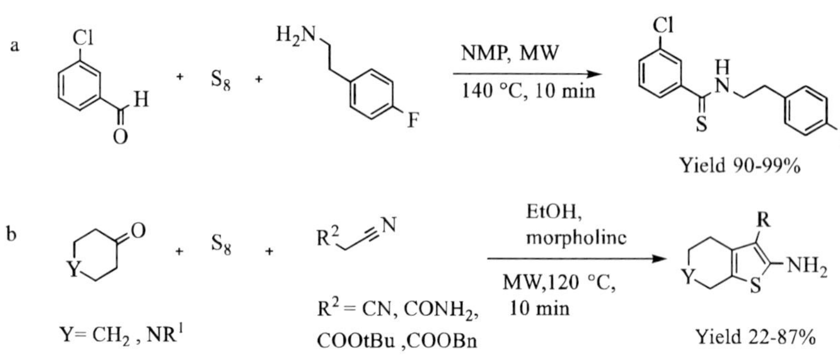

Scheme 39(a, b). Condensation reaction involving elemental sulphur.

The Gewald synthesis, which produces benzothiophenes by condensation of a ketone with an acyl ester in the presence of elemental sulfur, is somewhat related to this process (Scheme 39-b). Important structural motifs in natural chemicals, pharmaceutically active components, and dyes include fused thiophenes. The Gewald synthesis, like the Biginelli reaction, is well-suited to parallel library production since different building blocks can be processed under the same reaction conditions. (Treu et al. 2008).

The classical Pechmann approach for the synthesis of coumarins *via* the MW-promoted reaction (Singh et al. 1998; Bogdal 1998) (Scheme 40) has been extended to a solvent-free system wherein salicylaldehydes undergo Knoevenagel condensation with a variety of ethyl acetate derivatives in the presence of piperidine to afford coumarins.

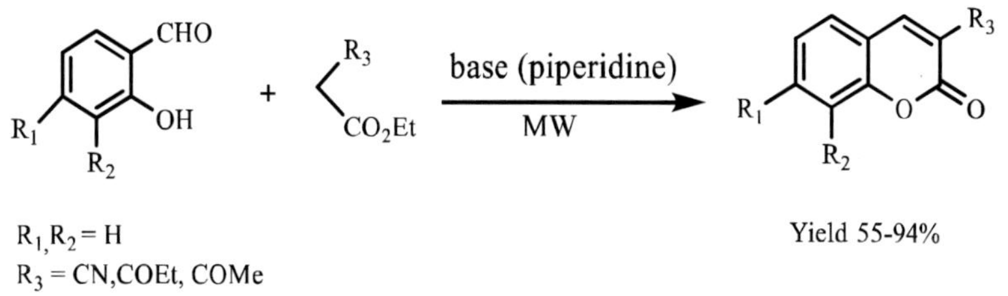

Scheme 40. MW-assisted synthesis of coumarins.

Addition Reactions

Similar to condensations, addition reactions are frequently used because they enable the formation of many C-C and C-heteroatom bonds. As an example, since high temperatures are usually required to maintain tolerably fast reaction durations, microwave-mediated techniques are ideal for Michael's additions to introduce anilines in heterocycle synthesis. Isatin, β-diketones, and isoxazoles were combined in an intriguing method to create fused spirooxindoles without the use of a catalyst or a solvent. The intermediate proceeds through a Michael addition with the isoxazole following an initial Knoevenagel condensation, and a subsequent intramolecular cyclocondensation produces the required spiro products (Yuvaraj, Manivannan, and Reddy 2015) (Scheme 41).

The many cycloadditions used to create useful heterocyclic molecules are even more common. Under a variety of circumstances, particularly Diels-Alder cycloadditions to produce significant 6-membered ring structures are carried out. For this transition, a wide range of solvents as well as numerous dienes and dienophiles can be used.

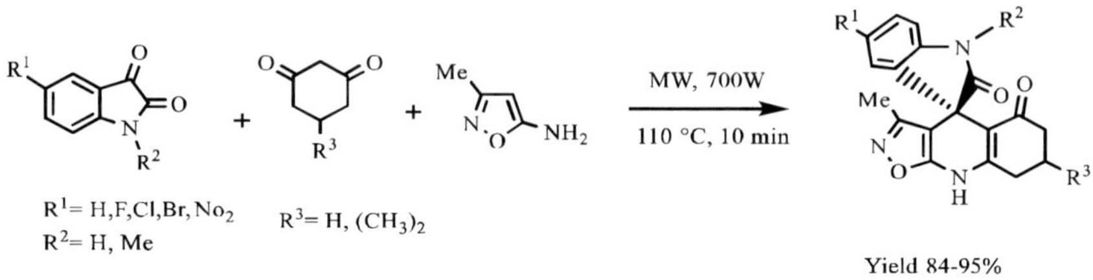

Scheme 41. Tandem Knovenagel/Michael addition reaction.

The example reaction shown in (Scheme 42) that even reactions involving substantial volumes of solid components can be carried out effectively (Razzaq and Kappe 2008).

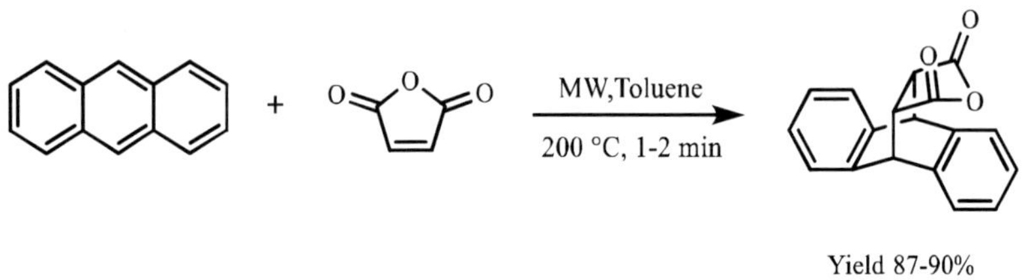

Scheme 42. Diels-alder cycloaddition of anthracene with maleic anhydride.

Substitution Reaction

Under microwave irradiation, the nucleophilic acyl substitution reaction of 5-aminouracil with ethyl cyanoacetate produced the corresponding 2-cyano-N-(2,4-dioxo-1,2,3,4-tetrahydropyrimidin5-yl)acetamide in excellent yield as shown in the following scheme. (Fahim, Shalaby, and Ibrahim 2019) (Scheme 43).

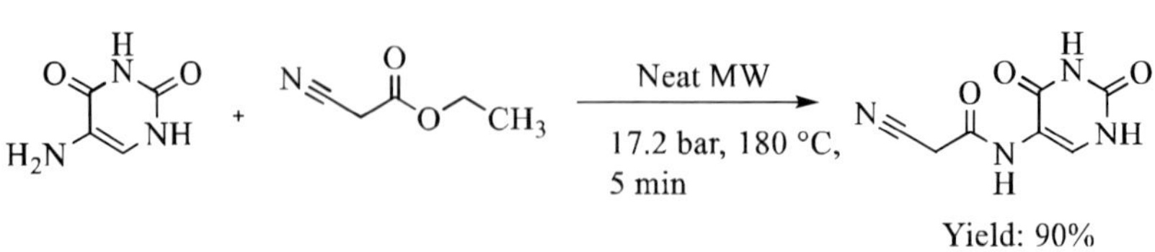

Scheme 43. Nucleophilic acyl substitution reaction.

The radiolabeling of molecules is a particular version of a substitution process. The exchange of iodine with the radioisotope ^{131}I to create radiolabelled hippurate for cancer treatment was demonstrated by a research team from Taiwan (Sherlock Huang et al. 2014) (Scheme 44). Standard sodium iodide and o-iodohippurate were combined using the practical microwave technique, and after 30 minutes, good radiochemical purities >90% were produced. In addition to saving time over traditional reflux settings, this microwave-mediated approach also significantly reduces the catalyst amount.

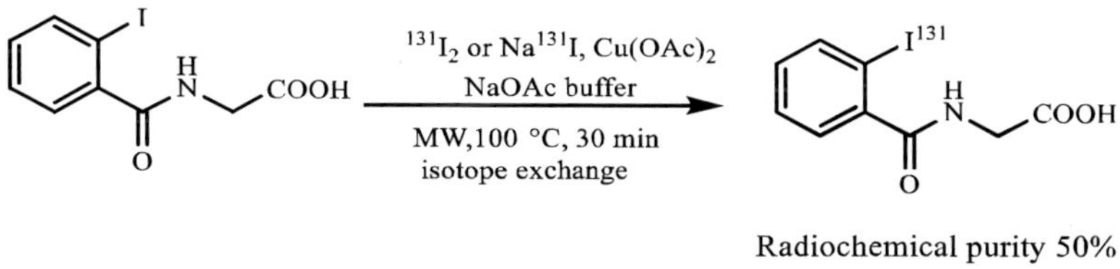

Scheme 44. Isotope exchange under microwave irradiation.

Singh and co-workers introduced a catalyst-free transamidation reaction in dioxane, which can be considered a nucleophilic carbonyl substitution altering the amide nitrogen (Vanjari, Allam, and Singh 2013) (Scheme 45). To create 24 architecturally intriguing carboxamides, several fundamental aryl amines were combined with formamide and acetamide. Multiple functionalities with both electron-donating and electron-withdrawing groups are tolerated by the mild yet effective reaction conditions. However, the substrate concentration and vial filling capacity must be regulated within the operational parameters of the reactor being used since volatile ammonia is produced as a byproduct.

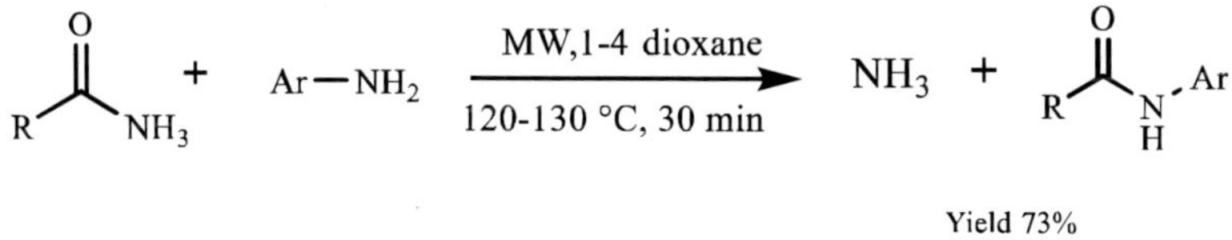

Scheme 45. Catalyst free transamidation reaction in dioxane.

The preparation of diphenyl ethers (Frlan and Kikelj 2006) (Scheme 46) as an important structural motif in pharmacologically active compounds *via* intermolecular nucleophilic aromatic substitution (S_NAr) has certain industrial relevance. An easy method of creating diaryl ethers using microwave energy with electron-poor phenols and aryl halides. Diaryl ethers produced a good coupling process and high yields (74–98%) according to the phenol's superior coupling tolerance.

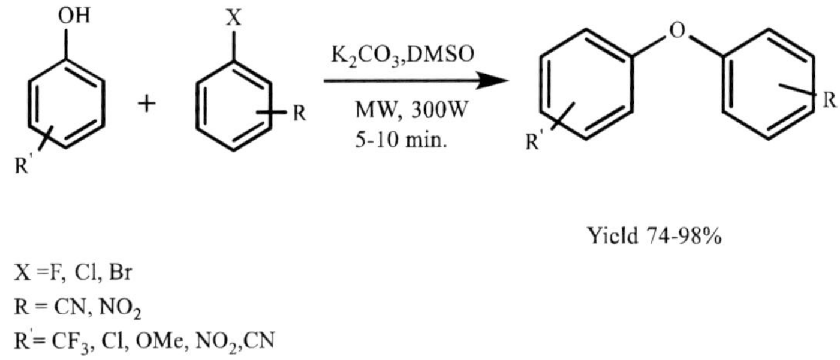

Scheme 46. Preparation of diphenyl ethers *via* intermolecular nucleophilic aromatic substitution.

Substitution reactions are prevalent in synthetic chemistry; they are routinely conducted under microwave irradiation, just like addition processes. Such reactions result in a monosubstituted chloropyrimidine derivative of pyrimidine scaffolds when one of the chlorine atoms is substituted by an amine or (hetero)aryl moiety (Hartung et al. 2006; Hartwig J F.2006) (Scheme-47).

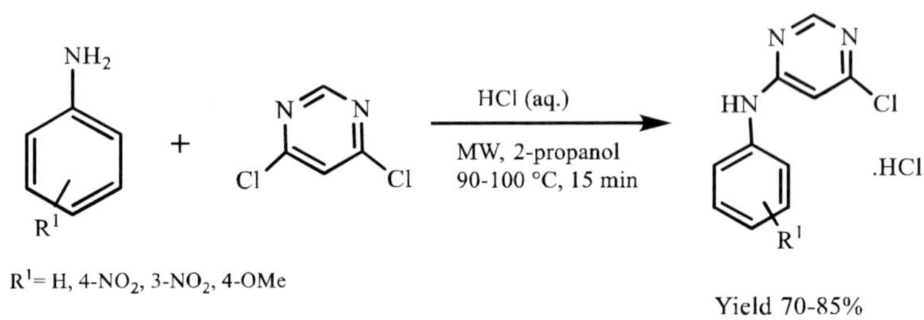

Scheme 47. Amination of pyrimidine scaffolds.

The same study demonstrates the benefit of microwave reactors with operation limitations >20 bar by using ammonia for the amination of chloropyrimidine derivatives *via* S_NAr (Hartung et al. 2006) (Scheme 48). The increased pressure limit allows you to process more volumes in a single batch than instruments with a 20-bar limit, which boosts productivity.

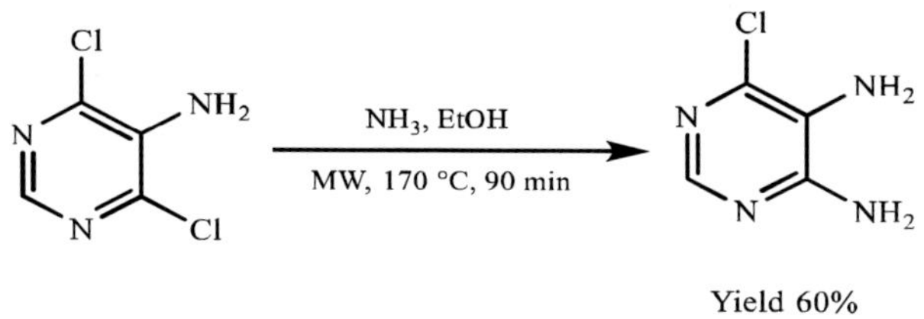

Scheme 48. Amination of pyrimidines.

Rearrangements

Due to high activation barriers, rearrangements are somewhat tricky reactions that require harsh conditions and long reaction times even under sealed-vessel conditions.

The compound (allyloxy)benzene was selected as a model substrate for the Claisen rearrangement studies. The substrate was reacted at a predetermined temperature in commercially available organic solvents under microwave irradiation in a sealed vial. The findings are shown in a (Scheme 49). To our delight, the reaction was noticeably sped up by the usage of PMA. After two minutes at 300°C and with a small quantity of PMA, the reaction produced a good yield (Hui et al. 2020; Velappan et al. 2019).

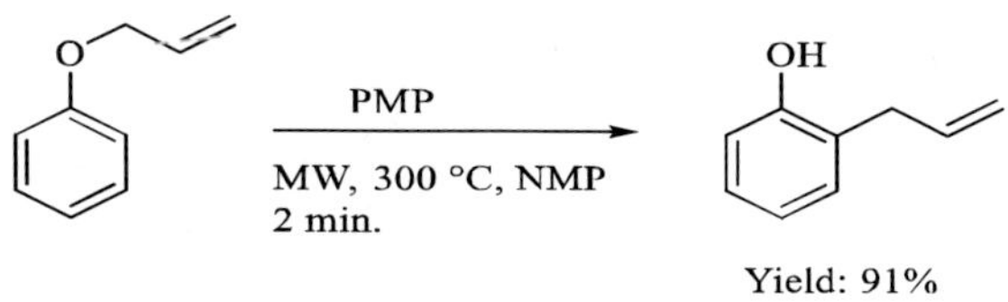

Scheme 49. Microwave-assisted Claisen rearrangement.

The creation of C-N bonds has found extensive use in [3,3]-Sigmatropic rearrangements, such as the rearrangement of allylic cyanates or allyloxy iminodiazaphospholidines. Potassium carbonate and ortho-dichlorobenzene were used to provide ideal conditions (Scheme 50-a,b). The doubly rearranged products were formed as single regio- and diastereomers (95:5) after being exposed to controlled microwave radiations for 15 minutes at a temperature of 180°C with overall good yields (Tjeng, Handore, and Batey 2020).

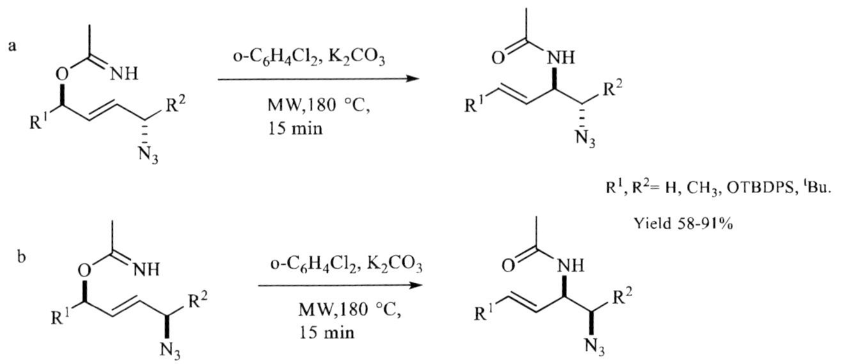

Scheme 50(a, b). A domino [3,3]-sigmatropic (Winstein and Overman) rearrangement.

An interesting approach was presented by Schmidt and co-workers to prepare allyl- and prenyl coumarins in a one-pot tandem Claisen rearrangement/Wittig Olefination/cyclization sequence (Schultze and Schmidt 2018). A vessel designed for microwave irradiation was filled with a solution of allyl ether and ylide in N, N-diethylaniline.

The substrate range could be expanded to include several highly substituted 2H-chromones with biological activity by using high-temperature sealed-vessel conditions. By heating the reacted N, N-diethylaniline over its boiling point, reaction times to complete the conversion might be shortened from up to 12 hours of overnight processing to just 1 hour (Scheme 51).

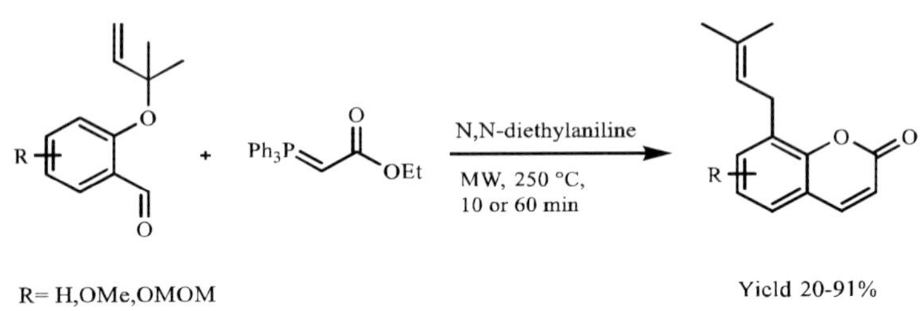

Scheme 51. Tandem Claisen rearrangement/Witting Olefination/cyclization to form coumarins.

A typical example often used in the pharmaceutical industry is the Newman-Kwart rearrangement of thiocarbamates (Moseley et al. 2006) (Scheme 52) Reaction temperatures are considerably above 200°C and process periods depending on the substrates. The reaction operates in a range of scales and on numerous instruments. Furthermore, microwave technology has shown to be incredibly practical in a lab setting for achieving the high temperatures needed (200–300°C), both for the chemist's safety and for quick reactions.

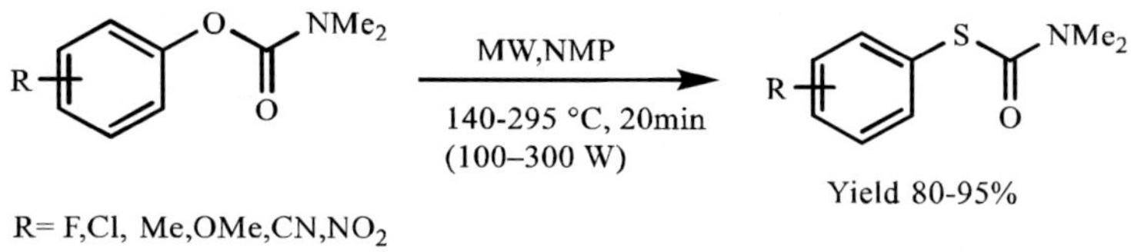

Scheme 52. Newman-Kwart rearrangement of thiocarbamates.

Under high-temperature sealed-vessel circumstances, several additional transformations of a comparable nature, such as Wolff, Hoffmann, or Pinacol rearrangements, can also be accomplished in MW irradiation.

Conclusion and Future Aspects

Microwave-assisted synthesis is a convenient way towards the goal of green chemistry. Ideally, the use of green chemistry ideas and techniques results in environmental regulation, control, clean up, and remediation. Therefore, their advantages can be described in terms of their economic effects. One of the many techniques used under the umbrella term of green chemistry is a microwave irradiation. Drug development is a difficult and demanding symphony that no one technique can whistle alone. To play it well, we need to use an entire orchestra of tactics and methods. Together, we are stronger. There has been a growing interest in using microwave irradiation in organic synthesis. Microwave heating has proved to be a valuable tool for synthetic chemists because it can improve the yields of the products and enhance the reaction rate. It is also safe and convenient for heating reaction mixtures at high temperatures. Nowadays, scientists are getting interested in synthesizing practically valuable complex scaffolds by using a microwave-initiated technique. Microwave-initiated protocols generally have various green features, including solvent-free reaction, atom economy, purification without chromatography, and high yields. There is a great expectation that the microwave process will find its place in industries because of its greenness, and time is not so far to replace many conventional technologies of synthesis of this emerging technology. Microwave chemistry has long been combined with other enabling technologies and strategies, such as multi-component reactions, solid-phase organic synthesis, and combinatorial chemistry. The combination of a multidisciplinary approach and microwave heating encourages scientists to explore previously unknown areas of complex pharmaceutical systems. To meet the increased synthetic demands in industry and academia, microwave irradiation still needs to find promising and enticing combinations with more techniques Overall, microwave heating is unquestionably a boon to organic and medicinal chemistry researchers. This enabling technology has progressed from the "last sort" to the "first choice" for carrying out synthetic transformations requiring heat.

In the future, with lower costs, microwave synthesizers will become an integral part of and a standard technology in most synthetic laboratories and they will continue to have a significant impact on both microwave assisted organic synthesis and drug discovery. The future of microwave technology looks bright because of its efficiency and its potential to contribute to clean products. Last but not least, the mechanism of each microwave-assisted coupling reaction should be evaluated with extreme care, as proper knowledge of the mechanism can drive researchers to construct new methods. We hope that this chapter would encourage the young researchers of chemistry as well as the specialists of green and sustainable chemistry.

References

Abenhaï M., David, André Loupy, Chu Pham, Ngoc Son, and Nguyen B. A. Hiep. 1994. "Synthesis of Jasminaldehyde by Solid-Liquid Phase Transfer Catalysis without Solvent, under MicrowaveIrradiation." *Synthetic Communications* 24 (9): 1199–1205. https://doi.org/10.1080/00397919408011718.

Alacid, Emilio, and Carmen Nájera. 2008. "Aqueous Sodium Hydroxide Promoted Cross-Coupling Reactions of Alkenyltrialkoxysilanes under Ligand-Free Conditions." *Journal of Organic Chemistry* 73 (6): 2315–22. https://doi.org/10.1021/jo702570q.

Allam, Bharat Kumar, and Krishna Nand Singh. 2011. "An Efficient Phosphine-Free Heck Reaction in Water Using Pd(l -Proline)2 as the Catalyst under Microwave Irradiation." *Synthesis*, no. 7: 1125–31. https://doi.org/10.1055/s-0030-1258452.

Anastas, P. T., J. C. Warne. 1998. "P. T. Anastas, J. C. Warne, Oxford University Press, Oxford, 2005th Edn, 1998;" *Oxford University Press, Oxford* 2005th edn.

Anastas, Paul T., and Mary M. Kirchhoff. 2002. "Origins, Current Status, and Future Challenges of Green Chemistry." *Accounts of Chemical Research* 35 (9): 686–94. https://doi.org/10.1021/ar010065m.

Arnáiz, Francisco J., and Rafael Aguado. 1995. "The Microscale Laboratory." *Journal of Chemical Education* 72 (10): 4–5. https://doi.org/10.1021/ed072pa196.

Bassyouni, Fatma A., Sherifa M. Abu-Bakr, and Mohamed Abdel Rehim. 2012. "Evolution of Microwave Irradiation and Its Application in Green Chemistry and Biosciences." *Research on Chemical Intermediates* 38 (2): 283–322. https://doi.org/10.1007/s11164-011-0348-1.

Beccalli, Egle M., Gianluigi Broggini, Michela Martinelli, and Silvia Sottocornola. "C−C, C−O, C−N Bond Formation on Sp2 Carbon by Pd(Ii)-Catalyzed Reactions Involving Oxidant Agents." *Chemical Reviews* 107, no. 11 (2007/11/01 2007): 5318-65. https://doi.org/10.1021/cr068006f.

Belwal, Savita. 2014. "*Green Revolution in Chemistry by Microwave Assisted Synthesis : A Review*" 1 (3): 22–25. https://doi.org/10.11648/j.mc.20130103.11.

Bernard, L., A. Friis-christensen, H.Pundt. 2008. *The European Information Society*. Edited by Michael P. Peterson William Cartwright, Georg Gartner, Liqiu Meng. Lecture No. Germany, Italy: Springer Berlin Heidelberg. https://www.ptonline.com/articles/how-to-get-better-mfi-results.

Bogdal, Dariusz. 1998. "Coumarins: Fast Synthesis by Knoevenagel Condensation under Microwave Irradiation." *Journal of Chemical Research - Part S*, no. 8: 468–69. https://doi.org/10.1039/a801724g.

Boger, Dale L., Susumu Miyazaki, Seong Heon Kim, Jason H. Wu, Steven L. Castle, Olivier Loiseleur, and Qing Jin. "Total Synthesis of the Vancomycin Aglycon." *Journal of the American Chemical Society* 121, no. 43 (1999/11/01 1999): 10004-11. https://doi.org/10.1021/ja992577q.

Bruckmann, Angelika, Anke Krebs, and Carsten Bolm. 2008. "Organocatalytic Reactions: Effects of Ball Milling, Microwave and Ultrasound Irradiation." *Green Chemistry* 10 (11): 1131–41. https://doi.org/10.1039/b812536h.

Chatel, Gregory, and Rajender S. Varma. 2019. "Ultrasound and Microwave Irradiation: Contributions of Alternative Physicochemical Activation Methods to Green Chemistry." *Green Chemistry* 21 (22): 6043–50. https://doi.org/10.1039/c9gc02534k.

Chem, S. 1979. "Stereoselective Synthesis of Arylated (E) -Alkenes by the Reaction of Alk-1 -Enylboranes with Aryl Halides in the Presence of Palladium Catalyst By." *J.C.S. Chem. Comm.*, no. 866: 866–67.

Chen, Gong, Jianwei Xie, Jiang Weng, and Xinhai Zhu. 2010. "Synthetic Communications : An International Journal for Rapid Communication of Synthetic Organic Chemistry CuI/PPh 3/PEG – Water : An Efficient Catalytic System for Cross-Coupling Reaction of Aryl Iodides and Alkynes." *Synthetic Communications*, no. October 2014: 37–41. https://doi.org/10.1080/00397911.2010.517363.

Chinchilla, Rafael, and Carmen Nájera. "The Sonogashira Reaction: A Booming Methodology in Synthetic Organic Chemistry." *Chemical Reviews* 107, no. 3 (2007/03/01 2007): 874-922. https://doi.org/10.1021/cr050992x.

Cívicos, José F, Diego A. Alonso, and Carmen Nájera. 2013. "Microwave-Promoted Copper-Free Sonogashira-Hagihara Couplings of Aryl Imidazolylsulfonates in Water." *Adv. Synth. Catal.* 355 (Scheme 1): 203–8. https://doi.org/10.1002/adsc.201.

Clark, James H., and Stewart J. Tavener. 2007. "Alternative Solvents: Shades of Green." *Organic Process Research and Development* 11 (1): 149–55. https://doi.org/10.1021/op060160g.

Clarke, Coby J., Wei Chien Tu, Oliver Levers, Andreas Bröhl, and Jason P. Hallett. 2018. "Green and Sustainable Solvents in Chemical Processes." *Chemical Reviews* 118 (2): 747–800. https://doi.org/10.1021/acs.chemrev.7b00571.

Clarke, Matthew L. "First Microwave-Accelerated Hiyama Coupling of Aryl- and Vinylsiloxane Derivatives: Clean Cross-Coupling of Aryl Chlorides within Minutes." *Advanced Synthesis & Catalysis* 347, no. 2-3 (2005/02/01 2005): 303-07. https://doi.org/10.1002/adsc.200404196.

Cordovilla, Carlos, Camino Bartolomé, Jesús Ma Martínez-Ilarduya, and Pablo Espinet. 2015. "The Stille Reaction, 38 Years Later." *ACS Catalysis* 5 (5): 3040–53. https://doi.org/10.1021/acscatal.5b00448.

Cravotto, Giancarlo, Emanuela Calcio Gaudino, Silvia Tagliapietra, Diego Carnaroglio, and Antonio Procopio. 2012. "A Green Approach to Heterogeneous Catalysis Using Ligand-Free Metal-Loaded Cross-Linked Cyclodextrins." *Green Processing and Synthesis* 1 (3): 269–73. https://doi.org/10.1515/gps-2012-0029.

Desai, Bimbisar, and C. Oliver Kappe. "Microwave-Assisted Synthesis Involving Immobilized Catalysts." *Immobilized Catalysts: Solid Phases, Immobilization and Applications* (2004): 177-208.https://doi.org/10.1007/b96876.

Doucet, Henri, and Jean-cyrille Hierso. 2007. "*Palladium-Based Catalytic Systems for the Synthesis of Conjugated Enynes by Sonogashira Reactions and Related Alkynylations Angewandte.*" 834–71. https://doi.org/10.1002/anie.200602761.

Elangovan, Arumugasamy, Yu-hsiang Wang, and Tong-ing Ho. 2003. "*Sonogashira Coupling Reaction with Diminished Homocoupling.*" no. 5: 4371–74.

Engel, Robert. "Phosphonates as Analogues of Natural Phosphates." *Chemical Reviews* 77, no. 3 (1977): 349-67.

Erdélyi, Máté, and Adolf Gogoll. "Rapid Microwave Promoted Sonogashira Coupling Reactions on Solid Phase." *The Journal of Organic Chemistry* 68, no. 16 (2003/08/01 2003): 6431-34. https://doi.org/10.1021/jo034284s.

Erythropel, Hanno C., Julie B. Zimmerman, Tamara M. De Winter, Laurène Petitjean, Fjodor Melnikov, Chun Ho Lam, Amanda W. Lounsbury, Karolina E. Mellor, Nina Z. Janković, Qingshi Tu, Lauren N. Pincus, Mark M. Falinski, Wenbo Shi, Philip Coish, Desirée L. Plata and Paul T. Anastas. 2018. "The Green ChemisTREE: 20 Years after Taking Root with the 12 Principles." *Green Chemistry* 20 (9): 1929–61. https://doi.org/10.1039/c8gc00482j.

Escudié, Jean, Henri Ranaivonjatovo, and Leslie Rigon. "Heavy Allenes and Cumulenes Ece' and Ecce' (E = P, as, Si, Ge, Sn; E' = C, N, P, as, O, S)." *Chemical Reviews* 100, no. 10 (2000/10/01 2000): 3639-96. https://doi.org/10.1021/cr990013z.

European Commission. 2001. "White Paper, Strategy for a Future Chemicals Policy." *Office for Official Publications of the European Communities.*

Evano, Gwilherm, Nicolas Blanchard, and Mathieu Toumi. "Copper-Mediated Coupling Reactions and Their Applications in Natural Products and Designed Biomolecules Synthesis." *Chemical Reviews* 108, no. 8 (2008/08/01 2008): 3054-131. https://doi.org/10.1021/cr8002505.

Fahim, Asmaa M., Mona A. Shalaby, and Medhat A. Ibrahim. 2019. "Microwave-Assisted Synthesis of Novel 5-Aminouracil-Based Compound with DFT Calculations."

Journal of Molecular Structure 1194: 211–26. https://doi.org/10.1016/j.molstruc.2019.04.078.

Frlan, Rok, and Danijel Kikelj. 2006. "Recent Progress in Diaryl Ether Synthesis." *Synthesis*, no. 14: 2271–85. https://doi.org/10.1055/s-2006-942440.

Gabano, Elisabetta, and Mauro Ravera. 2022. "Microwave-Assisted Synthesis: Can Transition Metal Complexes Take Advantage of This 'Green' Method?" *Molecules* 27 (13): 1–38. https://doi.org/10.3390/molecules27134249.

Gabriel, Camelia, Sami Gabriel, Edward H. Grant, Ben S. J. Halstead, and D. Michael P Mingos. 1998. "Dielectric Parameters Relevant to Microwave Dielectric Heating." *Chemical Society Reviews* 27 (3): 213–23. https://doi.org/10.1039/a827213z.

Gallagher-Duval, Shawn, Gwénaëlle Hervé, Guillaume Sartori, Gérald Enderlin, and Christophe Len. 2013. "Improved Microwave-Assisted Ligand-Free Suzuki-Miyaura Cross-Coupling of 5-Iodo-2′-Deoxyuridine in Pure Water." *New Journal of Chemistry* 37 (7): 1989–95. https://doi.org/10.1039/c3nj00174a.

Gawande, Manoj B., Vasco D. B. Bonifácio, Rafael Luque, Paula S. Branco, and Rajender S. Varma. 2013. "Benign by Design: Catalyst-Free in-Water, on-Water Green Chemical Methodologies in Organic Synthesis." *Chemical Society Reviews* 42 (12): 5522–51. https://doi.org/10.1039/c3cs60025d.

Gedye, Richard, Frank Smith, Kenneth Westaway, Humera Ali, Lorraine Baldisera, Lena Laberge, and John Rousell. 1986. "The Use of Microwave Ovens for Rapid Organic Synthesis." *Tetrahedron Letters* 27 (3): 279–82. https://doi.org/10.1016/S0040-4039(00)83996-9.

Genov, Miroslav, Antonio Almorín, and Pablo Espinet. 2007. "Microwave Assisted Asymmetric Suzuki-Miyaura and Negishi Cross-Coupling Reactions: Synthesis of Chiral Binaphthalenes." *Tetrahedron Asymmetry* 18 (5): 625–27. https://doi.org/10.1016/j.tetasy.2007.03.001.

Giguere, Raymond J., Terry L. Bray, Scott M. Duncan, and George Majetich. 1986. "Application of Commercial Microwave Ovens to Organic Synthesis." *Tetrahedron Letters* 27 (41): 4945–48. https://doi.org/10.1016/S0040-4039(00)85103-5.

Girard, Simon A., Xiong Hu, Thomas Knauber, Feng Zhou, Marc-olivier Simon, Guo-jun Deng, and Chao-jun Li. 2012. "*Pd-Catalyzed Synthesis of Aryl Amines via Oxidative Aromatization of Cyclic Ketones and Amines with Molecular Oxygen*," no. 10: 9098–9101.

Hajipour, Abdol R., Kazem Karami, and Fatemeh Rafiee. 2012. "An Efficient Stille Cross-Coupling Reaction Catalyzed by Ortho-Palladated Complex of Tribenzylamine under Microwave Irradiation." *Applied Organometallic Chemistry* 26 (1): 27–31. https://doi.org/10.1002/aoc.1860.

Hartung, Christian G., Alexander C. Backes, Beatrice Felber, Andrea Missio, and Alexander Philipp. 2006. "Efficient Microwave-Assisted Synthesis of Highly Functionalized Pyrimidine Derivatives." *Tetrahedron* 62 (43): 10055–64. https://doi.org/10.1016/j.tet.2006.08.065.

Hartwig, John F. "Discovery and Understanding of Transition-Metal-Catalyzed Aromatic Substitution Reactions." *Synlett* 2006, no. 09 (2006): 1283-94.https://doi.org/10.1055/s-2006-939728.

Hassan, Jwanro, Marc Se, Christel Gozzi, Emmanuelle Schulz, and Marc Lemaire. 2002. "*Aryl – Aryl Bond Formation One Century after the Discovery of the Ullmann Reaction.*"

Heck, K. F., and J. P. Nolley. 1972. "Palladium-Catalyzed Vinylic Hydrogen Substitu-tion Reactions with Aryl, Benzyl, and Styryl Halides." *Journal of Organic Chemistry* 37 (14): 2320–22. https://doi.org/10.1021/jo00979a024.

Himmelspach, Alexander, Guido J. Reiss, and Maik Finze. 2012. "Microwave-Assisted Kumada-Type Cross-Coupling Reactions of Iodinated Carba-Closo-Dodecaborate Anions." *Inorganic Chemistry* 51 (4): 2679–88. https://doi.org/10.1021/ic202638k.

Hooshmand, Seyyed Emad, Bahareh Heidari, Roya Sedghi, and Rajender S. Varma. 2019. "Recent Advances in the Suzuki-Miyaura Cross-Coupling Reaction Using Efficient Catalysts in Eco-Friendly Media." *Green Chemistry* 21 (3): 381–405. https://doi.org/10.1039/c8gc02860e.

Horváth, István T. 2018. "Introduction: Sustainable Chemistry." *Chemical Reviews* 118 (2): 369–71. https://doi.org/10.1021/acs.chemrev.7b00721.

Hui, Zi, Songwei Jiang, Xiang Qi, Xiang Yang Ye, and Tian Xie. 2020. "Investigating the Microwave-Accelerated Claisen Rearrangement of Allyl Aryl Ethers: Scope of the Catalysts, Solvents, Temperatures, and Substrates." *Tetrahedron Letters* 61 (24): 151995. https://doi.org/10.1016/j.tetlet.2020.151995.

Hundertmark, Thomas, Adam F Littke, Stephen L Buchwald, and Gregory C Fu. 2000. "*Pd (PhCN) 2 Cl 2/P (t -Bu) 3 : A Versatile Catalyst for Sonogashira Reactions of Aryl Bromides at Room Temperature*," no. Scheme 1: 4–6.

Jablonkai, Erzsébet, and György Keglevich. 2015. "*Catalyst-Free P – C Coupling Reactions of Halobenzoic Acids and Secondary Phosphine Oxides under Microwave Irradiation in Water*" 56: 1638–40. https://doi.org/10.1016/j.tetlet.2015.02.015.

Jagtap, Sangeeta. 2017. "Heck Reaction—State of the Art." *Catalysts* 7 (9). https://doi.org/10.3390/catal7090267.

Jeevan Chakravarthy, A. S., Noor Shahina Begum, M. S. Krishnamurthy, and S. HariPrasad. 2019. "A Rapid Synthesis of 1-Chloro-2-Arylcyclohexenes Using MWAOS in Aqueous Media: Single Crystal Studies of Two Representative Compounds." *Molecular Crystals and Liquid Crystals* 682 (1): 65–76. https://doi.org/10.1080/15421406.2019.1655979.

Jiang, Yubo, and Shaojun Huang. 2013. "*A Microwave-Assisted Three-Component Synthesis of Arylaminomethyl Acetylenes : A Facile Access to Terminal Alkynes*," no. I: 407–10. https://doi.org/10.1055/s-0033-1340331.

Kantak, Abhishek A, Shathaverdhan Potavathri, Rose A Barham, Kaitlyn M Romano, and Brenton Deboef. 2011. "*Metal-Free Intermolecular Oxidative C À N Bond Formation via Tandem C À H and N À H Bond Functionalization*," 19960–65.

Kappe, C. Oliver, Alexander Stadler, and Doris Dallinger. *Microwaves in Organic and Medicinal Chemistry.* Vol. 52: John Wiley & Sons, 2012.

Kappe, C. Oliver, Doris Dallinger, and S. Shaun Murphree. 2009. *Kappe, C. Oliver, Doris Dallinger, and S. Shaun Murphree. Practical Microwave Synthesis for Organic Chemists. Wiley., 2009.* Edited by and S. Shaun Murphree Kappe, C. Oliver, Doris Dallinger. Wiley,. Wiley.

Kappe, C. Oliver, Doris Dallinger, Edited R. Mannhold, H. Kubinyi, G. Folkers, Dennis A. Allerton, Han Walker, K. Don, Martine J. Lira, and Sergio A. Leurs. n.d. *Methods and Principles in Medicinal Chemistry.* Previous Volumes of This Series: Pharmacokinetics and Metabolism in Drug Design Chemokine Receptors as Drug Targets Aspartic Acid Proteases as Therapeutic Targets Protein Kinases as Drug Targets Prodrugs And.

Karu, Ramesh, and Satyanarayana Gedu. 20 18. "Microwave Assisted Domino Heck Cyclization and Alkynylation: Synthesis of Alkyne Substituted Dihydrobenzofurans." *Green Chemistry* 20: 369. https://doi.org/10.1039/c7gc02606d.

Khan, Danish, Iram Parveen, and Saurabh Sharma. 2021. "*Design, Synthesis and Characterization of Aurone Based α, β - Unsaturated Carbonyl-Amino Ligands and Their Application in Microwave Assisted Suzuki, Heck and Buchwald Reactions.*" https://doi.org/10.1002/ajoc.202100638.

Kitanosono, Taku, Koichiro Masuda, Pengyu Xu, and Shu Kobayashi. 2018. "Catalytic Organic Reactions in Water toward Sustainable Society." *Chemical Reviews* 118 (2): 679–746. https://doi.org/10.1021/acs.chemrev.7b00417.

Knowles, Jonathan P., and Andrew Whiting. 2007. "The Heck-Mizoroki Cross-Coupling Reaction: A Mechanistic Perspective." *Organic and Biomolecular Chemistry* 5 (1): 31–44. https://doi.org/10.1039/b611547k.

Kokel, Anne, Christian Schäfer, and Béla Török. 2017. "Application of Microwave-Assisted Heterogeneous Catalysis in Sustainable Synthesis Design." *Green Chemistry* 19 (16): 3729–51. https://doi.org/10.1039/c7gc01393k.

Kumar, Anuj, Yun Kuang, Zheng Liang, and Xiaoming Sun. 2020. "Microwave Chemistry, Recent Advancements, and Eco-Friendly Microwave-Assisted Synthesis of Nanoarchitectures and Their Applications: A Review." *Materials Today Nano* 11: 100076. https://doi.org/10.1016/j.mtnano.2020.100076.

Lai, Dipti, Sumit Ghosh, and Alakananda Hajra. 2021. "*Biomolecular Chemistry,*" 4397–4428. https://doi.org/10.1039/d1ob00323b.

Larhed, M., C. Moberg, and A. Hallberg. 2002. *Larhed, M., C. Moberg, and A. Hallberg. "Acc. Chem. Res." (2002): 717-727.* Edited by and A. Hallberg. Larhed, M., C. Moberg. Acc. Chem. Acc. Chem. Res.

Larhed, Mats, and Peter Nilsson. 2005. "*Microwave Irradiation as a High-Speed Tool for Activation of Sluggish Aryl Chlorides in Grignard Reactions,*" 1596–1600. https://doi.org/10.1055/s-2005-869856.

Larhed, Mats, Anders Hallberg, and R F Org React. 1996. "*Coupling Reactions The Intermolecular Heck Reaction of Aryl Halides or Aryl Triflates with Olefins Is of the Utmost Importance in Substrates for Coupling Require Reaction Temperatures of 70 - 100°C 1e and Reaction Times Ranging from Hours to Days for Fu*" 3263 (16): 9582–84.

Larhed, Mats. 1996. "*Rapid Microwave-Assisted Suzuki Coupling on Solid-Phase*" 37 (45): 8219–22.

Laskar, Ranjini, Tanay Pal, Trisha Bhattacharya, Siddhartha Maiti, Munetaka Akita, and Debabrata Maiti. 2022. "Sustainable C-H Functionalization under Ball-Milling, Microwave-Irradiation and Aqueous Media." *Green Chemistry* 24 (6): 2296–2320. https://doi.org/10.1039/d1gc04530j.

Li, Chao-Jun, ed. 2012. *Handbook of Green Chemistry: Green Synthesis.Wiley-VCH, 2012.* Wiley-VCH Verlag.

Loupy, André, Alain Petit, Jack Hamelin, Françoise Texier-Boullet, Patrick Jacquault, and Daniel Mathé. 1998. "New Solvent-Free Organic Synthesis Using Focused Microwaves." *Synthesis* 5 (9): 1213–34. https://doi.org/10.1055/s-1998-6083.

Makhloutah, Aline, Danylo Hatych, Thomas Chartier, Lou Rocard, Antoine Goujon, François Xavier Felpin, and Piétrick Hudhomme. 2022. "An Investigation of Palladium-Catalyzed Stille-Type Cross-Coupling of Nitroarenes in Perylenediimide Series." *Organic and Biomolecular Chemistry* 20 (2): 362–65. https://doi.org/10.1039/d1ob02291a.

Marco, Bianca Aparecida de, Bárbara Saú Rechelo, Eliane Gandolpho Tótoli, Ana Carolina Kogawa, and Hérida Regina Nunes Salgado. 2019. "Evolution of Green Chemistry and Its Multidimensional Impacts: A Review." *Saudi Pharmaceutical Journal* 27 (1): 1–8. https://doi.org/10.1016/j.jsps.2018.07.011.

Martina, Katia, Maela Manzoli, Emanuela Calcio Gaudino, and Giancarlo Cravotto. 2017. "Eco-Friendly Physical Activation Methods for Suzuki–Miyaura Reactions." *Catalysts* 7 (4). https://doi.org/10.3390/catal7040098.

Martínez, Regina, Isidro M. Pastor, and Miguel Yus. 2014. "1,2-Functionalized Imidazoles as Palladium Ligands: An Efficient and Robust Catalytic System for the Fluorine-Free Hiyama Reaction." *European Journal of Organic Chemistry* 2014 (4): 872–77. https://doi.org/10.1002/ejoc.201301439.

Marzaro, Giovanni, Adriano Guiotto, and Adriana Chilin. 2009. "*Microwave-Promoted Mono- N -Alkylation of Aromatic Amines in Water : A New Efficient and Green Method for an Old and Problematic Reaction*," 774–76. https://doi.org/10.1039/b900750d.

Mehta, Vaibhav P., and Erik V. Van der Eycken. 2011. "Microwave-Assisted C–C Bond Forming Cross-Coupling Reactions: An Overview." *Chemical Society Reviews* 40 (10): 4925–36. https://doi.org/10.1039/c1cs15094d.

Mgbemena, Chinedum Ogonna, Danning Li, Meng Fang Lin, Paul Daniel Liddel, Kali Babu Katnam, Vijay Thakur Kumar, and Hamed Yazdani Nezhad. 2018. "Accelerated Microwave Curing of Fibre-Reinforced Thermoset Polymer Composites for Structural Applications: A Review of Scientific Challenges." *Composites Part A: Applied Science and Manufacturing* 115 (July): 88–103. https://doi.org/10.1016/j.compositesa.2018.09.012.

Mitra, Alok Kumar, Aparna De, and Nilay Karchaudhuri. 1999. "Solvent-Free Microwave Enhanced Knoevenagel Condensation of Ethyl Cyanoacetate with Aldehydes." *Synthetic Communications* 29 (16): 2731–39. https://doi.org/10.1080/00397919908086438.

Miyaura, Norio, and Akira Suzuki. 1995. "*Palladium-Catalyzed Cross-Coupling Reactions of Organoboron Compounds*," no. I: 2457–83.

Morgan, John P., Elizabeth E. Torres, Richard Averill, and Alex M. Brody. 2022. "Updating the Benzoin Condensation of Benzaldehyde Using Microwave-Assisted Organic Synthesis and N-Heterocyclic Carbene Catalysis." *Journal of Chemical Education*, 4–8. https://doi.org/10.1021/acs.jchemed.2c00982.

Moseley, Jonathan D., Rosalind F. Sankey, Olivier N. Tang, and John P. Gilday. 2006. "The Newman-Kwart Rearrangement Re-Evaluated by Microwave Synthesis." *Tetrahedron* 62 (19): 4685–89. https://doi.org/10.1016/j.tet.2005.12.063.

Motwani, Hitesh V., and Mats Larhed. 2013. "Diarylated Ethanones from Mo(CO)6-Mediated and Microwave-Assisted Palladium-Catalysed Carbonylative Negishi Cross-Couplings." *European Journal of Organic Chemistry*, no. 22: 4729–33. https://doi.org/10.1002/ejoc.201300610.

Mutule, Ilga, and Edgars Suna. 2005. "Arylzinc Species by Microwave Assisted Grignard Formation-Transmetallation Sequence: Application in the Negishi Coupling." *Tetrahedron* 61 (47): 11168–76. https://doi.org/10.1016/j.tet.2005.09.022.

Negishi, Ei-ichi, and Armin De Meijere, Eds. 2003. "*Handbook of Organopalladium Chemistry for Organic Synthesis*." In John Wiley & Sons.

Nüchter, M., B. Ondruschka, W. Bonrath, and A. Gum. 2004. "Microwave Assisted Synthesis – a Critical Technology Overview." *Green Chemistry* 6 (3): 128–41. https://doi.org/10.1039/b310502d.

Pal, Ashutosh, and Krishnanka Shekhar Gayen. 2021. "The Impact of Microwave Irradiation Reaction in Medicinal Chemistry: A Review." *Oriental Journal Of Chemistry* 37 (1): 01–24. https://doi.org/10.13005/ojc/370101.

Patel, Ashok Raj, Archana Asatkar, Geetika Patel, and Subhash Banerjee. 2019. "Synthesis of Rice Husk Derived Activated Mesoporous Carbon Immobilized Palladium Hybrid Nano-Catalyst for Ligand-Free Mizoroki-Heck/Suzuki/Sonogashira Cross-Coupling Reactions." *Chemistry Select* 4 (19): 5577–84. https://doi.org/10.1002/slct.201900384

Pii, Pergamon, Elsevier Science, Great Britain, Organic Chemistr, and Natural Sciences. 1996. "*Microwaves Assisted Gabriel Synthesis of Phthalides Margita Ldcov~i', Jarmila Chovancovd, Eva Veverkov~i and Stefan Toma*" 52 (47): 14995–6.

Polshettiwar, Vivek, and Rajender S. Varma. 2008. "Aqueous Microwave Chemistry: A Clean and Green Synthetic Tool for Rapid Drug Discovery." *Chemical Society Reviews* 37 (8): 1546–57. https://doi.org/10.1039/b716534j.

Prat, Denis, John Hayler, and Andy Wells. 2014. "A Survey of Solvent Selection Guides." *Green Chemistry* 16 (10): 4546–51. https://doi.org/10.1039/c4gc01149j.

Rahman, Matiur, Sumit Ghosh, Dhananjay Bhattacherjee, Grigory V. Zyryanov, Avik Kumar Bagdi, and Alakananda Hajra. 2022. "Recent Advances in Microwave-Assisted Cross-Coupling Reactions." *Asian Journal of Organic Chemistry* 11 (8). https://doi.org/10.1002/ajoc.202200179.

Ranu, Brindaban C., Raju Dey, Tanmay Chatterjee, and Sabir Ahammed. 2012. "Copper Nanoparticle-Catalyzed Carbon-Carbon and Carbon-Heteroatom Bond Formation with a Greener Perspective." *Chem Sus Chem* 5 (1): 22–44. https://doi.org/10.1002/cssc.201100348.

Rathi, Anuj K., Manoj B. Gawande, Radek Zboril, and Rajender S. Varma. 2015. "Microwave-Assisted Synthesis - Catalytic Applications in Aqueous Media." *Coordination Chemistry Reviews* 291: 68–94. https://doi.org/10.1016/j.ccr.2015.01.011.

Razzaq, Tahseen, and C. Oliver Kappe. 2008. "On the Energy Efficiency of Microwave-Assisted Organic Reactions." *Chem Sus Chem* 1 (1–2): 123–32. https://doi.org/10.1002/cssc.200700036.

Rogers, Luke, and Klavs F. Jensen. 2019. "Continuous Manufacturing-the Green Chemistry Promise?" *Green Chemistry* 21 (13): 3481–98. https://doi.org/10.1039/c9gc00773c.

Rummelt, Stephan M., Marco Ranocchiari, and Jeroen A. Van Bokhoven. 2012. "*Synthesis of Water-Soluble Phosphine Oxides by Pd/C-Catalyzed P À C Coupling in Water*," no. 9: 2003–5.

Saha, Pritam, and Koushik Mukherjee. 2022. "Sonogashira Reaction Cum Cyclization under Microwave Irradiation Supported by Sodium Aluminate (NaAlO2) and Catalyzed by Zinc: A Green Alternative Approach in Production of Furo[3,2-h]Quinolones." *Materials Today: Proceedings*, no. xxxx. https://doi.org/10.1016/j.matpr.2022.11.083.

Saikia, Pallabi, Gitarthi Sharma, Sanjib Gogoi, and Romesh C. Boruah. 2015. "Cascade Imination, Buchwald-Hartwig Cross Coupling and Cycloaddition Reaction: Synthesis of Pyrido[2,3-d]Pyrimidines." *RSC Advances* 5 (30): 23210–12. https://doi.org/10.1039/c5ra00725a.

Salih, Kifah S. M., and Younis Baqi. 2019. "Microwave-Assisted Palladium- Catalyzed Cross-Coupling Reactions : Generation of Carbon – Carbon Bond." *Catalysis* 10 (4): 35. https://doi.org/doi:10.3390/catal10010004.

Salih, Kifah S M. 2022. "*Modern Development in Copper- and Nickel-Catalyzed Cross-Coupling Reactions : Formation of Carbon-Carbon and Carbon- Heteroatom Bonds under Microwave Irradiation Conditions*" 202200023: 1–26. https://doi.org/10.1002/ajoc.202200023.

Sanghi, Rashmi. 2000. "Microwave Irradiation." *Resonance* 5 (3): 77–81. https://doi.org/10.1007/bf02839002.

Sarkar, Anirban, Sougata Santra, Shrishnu Kumar Kundu, Alakananda Hajra, Grigory V. Zyryanov, Oleg N. Chupakhin, Valery N. Charushin, and Adinath Majee. 2016. "A Decade Update on Solvent and Catalyst-Free Neat Organic Reactions: A Step Forward towards Sustainability." *Green Chemistry* 18 (16): 4475–4525. https://doi.org/10.1039/c6gc01279e.

Schruder, W. Christopher, Michael G. Organ, and William J. Pietro. 2016. "'Metal Nanoparticle Impregnated Controlled-Size Silica Macrospheres as a Microwave-Transparent Catalyst System for MACOS.' Current Nanoscience 12, No. 4 (2016): 448-454." *Current Nanoscience 12* 4: 448-454. https://www.ingentaconnect.com/content/ben/cnano/2016/00000012/00000004/art00009.

Schultze, Christiane, and Bernd Schmidt. 2018. "Prenylcoumarins in One or Two Steps by a Microwave-Promoted Tandem Claisen Rearrangement/Wittig Olefination/Cyclization Sequence." *Journal of Organic Chemistry* 83 (9): 5210–24. https://doi.org/10.1021/acs.joc.8b00667.

Shah, Dipen, and Harjinder Kaur. 2014. "Resin-Trapped Gold Nanoparticles: An Efficient Catalyst for Reduction of Nitro Compounds and Suzuki-Miyaura Coupling." *Journal of Molecular Catalysis A: Chemical* 381: 70–76. https://doi.org/10.1016/j.molcata.2013.10.004.

Sherlock Huang, Lin Chiang, Kou Hung Wu, Pi Wen Ko, Cheng Ying Hsieh, Kuan Chuan Pao, Shih Ching Chou, Fa Kuen Shieh, Radhakrishnan Sureshbabu, and Ming Hua

Hsu. 2014. "Microwave Assistance of Labeling Hippuric Acid by I-131." *Applied Radiation and Isotopes* 89: 53–57. https://doi.org/10.1016/j.apradiso.2014.02.006.

Singh, Jasvinder, Jasamrit Kaur, Sandeep Nayyar, and Goverdhan L. Kad. 1998. "Highly Efficient and Single Step Synthesis of 4-Phenylcoumarins and 3,4-Dihydro-4-Phenylcoumarins over Montmorillonite K-10 Clay, under Microwave Irradiation." *Journal of Chemical Research - Part S*, no. 5: 280–81. https://doi.org/10.1039/a708103k.

Sonogashira, Kenkichi, Yasuo Tohda, and Nobue Hagihara. "A Convenient Synthesis of Acetylenes: Catalytic Substitutions of Acetylenic Hydrogen with Bromoalkenes, Iodoarenes and Bromopyridines." *Tetrahedron letters* 16, no. 50 (1975): 4467-70.

Stadler, A., C. O. Kappe. 2001. "Microwave Assisted in Organic Chemistry." *J. Comb. Chem.* 3: 624.

Stadler, Alexander, Behrooz H. Yousefi, Doris Dallinger, Peter Walla, Erik Van Der Eycken, Nadya Kaval, and C. Oliver Kappe. 2003. "Scalability of Microwave-Assisted Organic Synthesis. From Single-Mode to Multimode Parallel Batch Reactors." *Organic Process Research and Development* 7 (5): 707–16. https://doi.org/10.1021/op034075+.

Strauss, Christopher R., and Robert W. Trainor. 1995. "Developments in Microwave-Assisted Organic Chemistry." *Australian Journal of Chemistry* 48 (10): 1665–92. https://doi.org/10.1071/CH9951665.

Suzuki, Akira. 1999. "*Recent Advances in the Cross-Coupling Reactions of Organoboron Derivatives with Organic Electrophiles, 1995 – 1998*" 576: 147–68.

Tanaka, K., and F. Toda. "Solvent-Free Organic Synthesis." *Chemical Reviews* 100, no. 3 (2000/03/01 2000): 1025-74. https://doi.org/10.1021/cr940089p

Taylor, Publisher, György Keglevich, Nóra Zsuzsa Kiss, Erzsébet Jablonkai, Erika Bálint, and Zoltán Mucsi. 2015. "*The Potential of Microwave in Organophosphorus Syntheses*," no. May: 37–41. https://doi.org/10.1080/10426507.2014.989430.

Tjeng, Andy A., Kishor L. Handore, and Robert A. Batey. 2020. "Stereocontrolled Microwave-Assisted Domino [3,3]-Sigmatropic Reactions: A Winstein-Overman Rearrangement for the Formation of Differentiated Contiguous C-N Bonds." *Organic Letters* 22 (8): 3050–55. https://doi.org/10.1021/acs.orglett.0c00801.

Toda, Fumio, Koichi Tanaka, Tomoyuki Tamashima, and Masako Kato. 1998. "*Stereoselective Thermal Conversion of via S- Cis -Diallene in the Crystalline State*" 630: 2724–27.

Treu, Matthias, Thomas Karner, Roland Kousek, Hehnut Berger, Moriz Mayer, Darryl B. McConnell, and Alexander Stadler. 2008. "Microwave-Assisted Parallel Synthesis of Fused Heterocycles in a Novel Parallel Multimode Reactor." *Journal of Combinatorial Chemistry* 10 (6): 863–68. https://doi.org/10.1021/cc800081b.

Valavanidis, Athanasios. 2016. "*Green Chemistry and New Technological Developments. New Avenues for the Green Economy and Sustainable Future of Science and Technology.*" Web-Site: Www.Chem.Uoa.Gr (Http://195.134.79.37/Scinews/Report/PDF/NEWS%20DEVELOPMENTS-GREEN-CHEM-PDF_33PAG-8-7-2016.Pdf) 1 (July): 1–33.

Vanjari, Rajeshwer, Bharat Kumar Allam, and Krishna Nand Singh. 2013. "A Novel and Simple Transamidation of Carboxamides in 1,4-Dioxane without a Catalyst." *Tetrahedron Letters* 54 (20): 2553–55. https://doi.org/10.1016/j.tetlet.2013.03.049.

Varma, Rajender S. "Greener and Sustainable Trends in Synthesis of Organics and Nanomaterials." *ACS Sustainable Chemistry & Engineering* 4, no. 11 (2016/11/07 2016): 5866-78. https://doi.org/10.1021/acssuschemeng.6b01623.

Varma, Rajender S. 2014. "Journey on Greener Pathways: From the Use of Alternate Energy Inputs and Benign Reaction Media to Sustainable Applications of Nano-Catalysts in Synthesis and Environmental Remediation." *Green Chemistry* 16 (4): 2027–41. https://doi.org/10.1039/c3gc42640h.

Velappan, Anand Babu, Subhashini Kogatam, Dhrubajyoti Datta, Rakshantha Srithar, Gunasekaran Nanjappan, and Joy Debnath. 2019. "2-Methoxyphenyl Isocyanate: A Chemoselective Multitasking Reagent for an Amine Protection/Deprotection Sequence." *Organic Chemistry Frontiers* 6 (14): 2360–64. https://doi.org/10.1039/c9qo00293f.

Vijayaraghavan, Reena, Ravindra P. Pal, and Ajayan Vinu. 2010. "*Ligand-Free Palladium-Catalyzed C – S Coupling Reactions Using Water As*," no. 18: 2733–36. https://doi.org/10.1055/s-0030-1258815.

Walla, Peter, and C. Oliver Kappe. 2004. "*Microwave-Assisted Negishi and Kumada Cross-Coupling Reactions of Aryl Chlorides †*," 564–65.

Xu, Guangli, Yuanmin Zhang, Kehu Wang, Ying Fu, and Zhengyin Du. 2015. "Microwave-Assisted Stille Cross-Coupling Reaction Catalysed by in Situ Formed Palladium Nanoparticles." *Journal of Chemical Research* 39 (7): 399–402. https://doi.org/10.3184/174751915X14357494708710.

Yuvaraj, Panneerselvam, Karthikeyan Manivannan, and Boreddy S. R. Reddy. 2015. "Microwave-Assisted Efficient and Highly Chemoselective Synthesis of Oxazolo[5,4-B]Quinoline-Fused Spirooxindoles via Catalyst- and Solvent-Free Three-Component Tandem Knoevenagel/Michael Addition Reaction." *Tetrahedron Letters* 56 (1): 78–81. https://doi.org/10.1016/j.tetlet.2014.11.001.

Zhu, X.-H., G. Chen, Y. Ma, H.-C. Song, Z.-L. Xu, & Y.-Q. Wan, (2007). "A General, Highly Efficient Ullmann C-O Coupling Reaction under Microwave Irradiation and the Effects of Water." *Chinese Journal of Chemistry,* 25 (04009718): 546–52.

Biographical Sketch

Premlata Kumari, PhD

Affiliation: Department of Chemistry, Sardar Vallabhbhai National Institute of Technology, Surat-395007 (Gujarat), India

Business Address: Department of Chemistry, Sardar Vallabhbhai National Institute of Technology, Surat-395007 (Gujarat), India

Research and Professional Experience: 16 Years

Professional Appointments: Associate Professor

Honors: *Best Researcher Award* for the Contribution and Honourable Achievement in Innovative Research by ScienceFather Scifax company Approved and Registered by Ministry of Corporate Affairs (MCA), Govt. of India.

Publications from the Last 3 Years:

(1) Ajayrajsinh R. Zala, Dhanji P. Rajani, Iqrar Ahmad, Harun Patel & Premlata Kumari (2023) Synthesis, characterization, molecular dynamic simulation, and biological assessment of cinnamates linked to imidazole/benzimidazole as a CYP51 inhibitor, *Journal of Biomolecular Structure and Dynamics,* DOI: 10.1080/07391102.2023.2170918

(2) Patel, K.B., Mukherjee, S., Bhatt, H., Rajani, D., Ahmad, I., Patel, H. and Kumari, P., 2022. Synthesis, docking, and biological investigations of new coumarin-piperazine hybrids as potential antibacterial and anticancer agents. *Journal of Molecular Structure*, p.134755.

(3) Zala, A.R., Rajani, D.P. and Kumari, P., 2022. Synthesis, molecular docking, ADME study, and antimicrobial potency of piperazine based cinnamic acid bearing coumarin moieties as a DNA gyrase inhibitor. *Journal of Biochemical and Molecular Toxicology*, p.e23231.

(4) Patel, K.B. and Kumari, P., 2022. A review: Structure-activity relationship and antibacterial activities of Quinoline based hybrids. *Journal of Molecular Structure*, p.133634.

(5) Zala, A.R., Rajani, D.P. and Kumari, P., 2022. Design, synthesis, molecular docking and antimicrobial and antimycobacterial activities of novel hybrid of coumarin-cinnamic acids. *Chemical Data Collections*, *39*, p.100862.

(6) Patel, K.B. and Kumari, P., 2022. Anticancer activity and docking study of flavone derivatives as peroxisome proliferator-activated receptorγ inhibitors. *Structural Chemistry*, pp.1-17.

(7) Azazahemad A. Kureshi, Jinal Patel, Chirag Dholakiya, Ashish Kar, Amit Mirgal, S.P. Salvi, Tabaruk Hussain, Premlata Kumari, Sharad K. Tripathi, Raghuraj Singh, Urja Gada, Satyanshu Kumar Simultaneous determination of five bioactive constituents in fruit rind extracts of *Garcinia lanceifolia* using a validated UHPLC-PDA method and

evaluation of antioxidant activity,(2021) *Medicinal Plants-International Journal of Phytomedicines and Related Industries,* 13(1) 56-65.

(8) Azazahemad A. Kureshi, Hiral M. Vaghela, Satyanshu Kumar, Raghuraj Singh, Premlata Kumari (2020) Green Synthesis of Gold Nanoparticles Mediated by Garcinia Fruits and Their Biological Applications. *Pharmaceutical Sciences* 27(2) 238-250.

(9) Azazahemad Kureshi, Amit Mirgal, Siddhesh Salvi, Premlata Kumari, Raghuraj Singh and Satyanshu Kumar, (2020) Antioxidant Activities and Phenolics Contents of *Garcinia talbotii* Fruit Rind, *Research Journal of Medicinal Plants,* 14 (3) 104-110

(10) Kureshi, A.A., Dholakiya, C., Hussain, T., Mirgal, A., Salvi, S.P., Barua, P.C., Talukdar, M., Beena, C., Kar, A., Zachariah, T.J., Kumari, P., Dhanani, T., Singh, R., & Kumar, S. (2020). Simultaneous identification and quantification of three biologically active xanthones in *Garcinia* species using a rapid UHPLC-PDA method. *Acta Chromatographica,* 32(3) 179-188.

Chapter 2

Plant Extract-Based Green Synthesis of Zinc Oxide Nanomaterials and Its Greener Prospects for the Future

Subhodeep Barman[1,2]
and Rahul Das[1,*]
[1]Department of Physics, The University of Burdwan, Golapbag, West Bengal, India
[2]Department of Physics, Gangarampur College, Dakshin Dinajpur, West Bengal, India

Abstract

Different unique properties of zinc oxide (ZnO) nanomaterials allow it to occupy a prominent position in modern technologies in various aspects. This chapter covers the green route synthesis of ZnO nanomaterials and details the future prospects of these nanomaterials. The green synthesis method has many advantages over the traditional physical and chemical synthesis methods in terms of more manageable, less time-consuming, inexpensive, non-toxic, environmentally friendly, reliable and sustainable development of modern society. Green synthesis belongs to the 'bottom-up' approach of nano-synthesis processes, similar to chemical reduction, in which reducing agents are replacing by plant extracts (containing the phytochemicals like polyphenols, flavonoids and terpenoids etc.) having inherent stabilizing and capping properties. During the green synthesis process, plant extracts are mixed in a zinc precursor solution. The precursors act as the source of zinc ions. Finally, the ions are reduced by the phytochemicals present in the extract and precipitates as zinc oxide nanomaterials. A series of experimental techniques are implemented for the usual confirmatory testing of the

[*] Corresponding Author's Email: rdas@phys.buruniv.ac.in.

In: The Future of Green Synthesis
Editor: Jacques Kent
ISBN: 979-8-88697-816-2

synthesized materials and results show that green synthesized zinc oxide nanomaterials can exhibit excellent antibacterial, anticancer, antioxidant and photocatalytic dye degradation etc. properties. In areas where purity is the primary concern, such as for example, in medical, biological, agricultural and food industries, the development of plant-based ZnO nanomaterials free from chemical impurities is required. From this point of view, detailed study is required to know the green mechanisms related to the principles of green chemistry and their activities during the plant-based synthesis of ZnO nanomaterials.

Introduction

Zinc and its compounds have always played an important role in our healthy ecosystem. However, the realization of this valuable position of zinc does not come suddenly. In 1869, when a student of Louis Pasteur named Raulin studied the growth of *Aspergillus niger*, the importance of zinc for the growth of microorganisms was first noted [1]. After a long time, Mendel and Bradley provided a clue to the importance of zinc to animals in 1905 when they reported on the constituents of a snail's respiratory pigment [2]. In 1926, when Sommer and Lipman observed the growth of sunflowers and barley, they realized the requirement of zinc for plant growth [3]. In 1934, Wilbert Todd and his colleagues at the University of Wisconsin discovered that zinc is a growth factor in rats. From this report, zinc was identified as an essential element for the growth of larger animals [4]. In 1958, the growth of poultry was shown to be influenced by zinc [5]. In 1963, Prasad et al. established for the first time the essential zinc requirement for human health [6]. The human body contains around 2-3 g of zinc. Almost 10-15 mg zinc is needed per day for a healthy human body. A deficiency of zinc in the human system can inactive the various enzymes like carbonic anhydrase, carboxypeptidase and alcohol dehydrogenase. In fact, without zinc, various physiological functions will be difficult to regulate. Due to the tiny size of nanoparticles, the required zinc can easily be absorbed by the human body from the zinc oxide nanomaterials which are used as food additive now-a-days [7]. From a nanoscale perspective, zinc oxide has many advantages that can revolutionize this modern society. Due to the increased surface-to-volume ratio, nano-sized ZnO exhibits a variety of enhanced physicochemical activities, including antibacterial, photocatalytic, antioxidant, drug delivery and anticancer etc. There are many physical and chemical techniques used for the fabrication of pure and doped ZnO nanomaterials. However, most of these methods require high pressure and high temperature to maintain harsh experimental conditions through many complicated and problematic steps in expensive modern facilities. Apart from this, almost all methods deal with various dangerous, health-endangering chemicals during the synthesis; adsorb the toxic chemicals that may not be suitable for medical applications and suffer from short stability [8-12]. The green nanoparticle synthesis process based on plant extracts has so many advancements (shown in Figure 1) such as eco-friendly, fast, impurity-free, non-toxic, easily handled reactants, simple reaction conditions, minimize energy inputs and after all, this is a cost-effective process [13, 14]. Plant extracts, in this case, act as both reducing and stabilizing agents. Different parts of plants such as leaves, roots, stems, flowers, fruits and seeds have been used as a source of natural reactants during the synthesis of nanomaterials [15].

Figure 1. Main advantages of green synthesis of ZnO nanomaterials based on plant-extracts.

Figure 2 represents the statistics of the number of published articles in the scientific literature over the last 22 years (2000 – 2022). In this case, we relied on the data from Scopus publications after using the keyword ‘green synthesis of ZnO’.

According to Figure 2 and the report by Huang et al. [16], it can be clarified that the number of scientific papers on 'green synthesis of ZnO' was almost irregular between 2003-2017 and then from 2018 to last year in 2022, the number increased considerably and this number is expected to increase further in coming years. Thus, it can at least be confirmed that the increasing trend indicates the growing interest of researchers in the green synthesis of ZnO. The following parts contains the detailed discussion on the synthesis of ZnO using plant extracts, the significant role of the phytochemicals including their mechanisms during the synthesis process along with their immense potential for a variety of realistic daily applications: antibacterial activity, drug delivery, anticancer agents, photocatalysis, solar cells, nano-fertilizers and their novel applications in cosmetic fields.

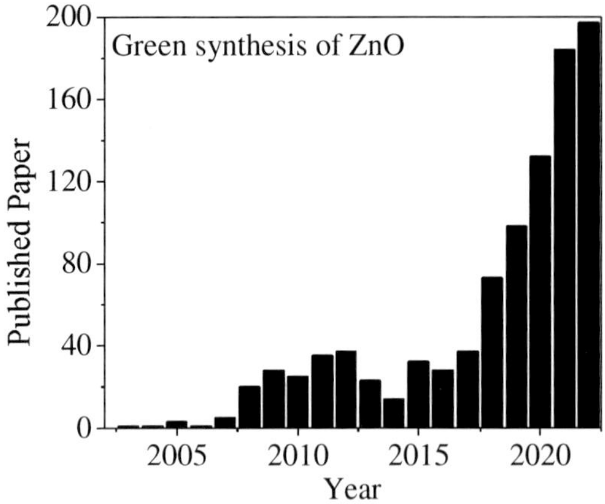

Figure 2. Statistics of the number of research articles published over the past twenty-two years (2000 – 2022).

ZnO Nanomaterial Synthesis

Using Traditional Methods

There are two traditional approaches referred to as 'bottom-up' and 'top-down' that are known for the synthesis of all types of nanomaterials as well as ZnO nanomaterials. The bottom-up approach is the assembly of basic units (e.g., atoms/ions/molecules) into larger structures through various chemical or physical techniques such as co-precipitation, hydrothermal, spray pyrolysis, wet chemistry and sol-gel etc. [17–19]. At the same time, the formation of nanomaterials, from the largest to the smallest, falls under the category of 'top-down' approach which is mainly related to physical processes such as pulsed laser ablation, metallurgical, molecular beam epitaxy, lithography, mechanical ball milling, exfoliation and so on [20–22]. Figure 3 helps to visualize traditional synthesis methods of ZnO nanomaterials using the top-down and bottom-up approaches. However, these traditional physical and chemical nano-synthesis methods are usually expensive, energy-consuming, labor-intensive, complicated in the sense of sophisticated infrastructures. Also, traditional methods of nano synthesis use hazardous chemicals and its byproducts are not suitable for a healthy environment. For these reasons mentioned above, we need the simplest, environment-friendly and cost-effective method not only for the synthesis of ZnO nanoparticles (ZnO-NPs), but also for any kind of nanomaterial synthesis.

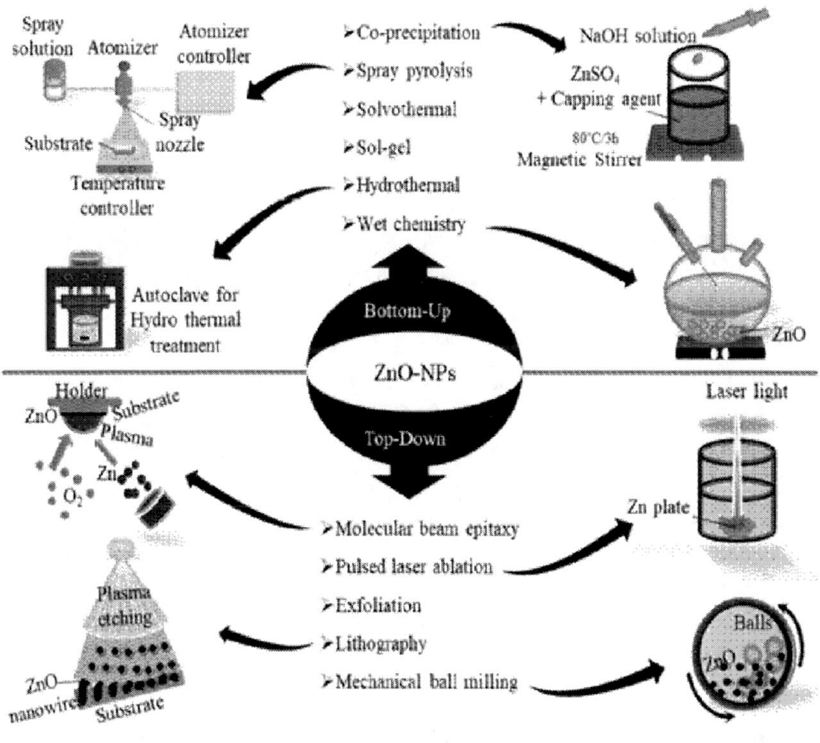

Figure 3. Different traditional methods for the synthesis of ZnO nanomaterials by top-down and bottom-up approaches.

Using Plant Extracts

To find alternatives of traditional methods, researchers have developed many biological methods that are environmentally safe and cost-efficient for the synthesis of nanomaterials. However, some microbes-based methods require extra care to preserve the bacterial cells and additional time for the bacterial/fungal cultures. On the other hand, the 'green' synthesis of nanomaterials based on plant extracts is a time-efficient, easy-to-maintain method and has so many advantages, which are shown in Figure 1. These plant-based synthesized nanomaterials remain stable for a long time. There are so many phytochemicals present in the extracts of various parts of plants such as: leaves, roots, stems, barks, flowers, fruits, seeds and peels etc. These phytochemicals act like natural reagents during the plant-based green synthesis process. The detailed analysis of the plant exacts and their correct role for the synthesis of ZnO nanoparticles is still under investigation. However, according to the available scientific reports, some active plant phytochemicals or biomolecules are believed to play an important role in the production of nanocrystalline ZnO. In fact, different parts of plants contain different types of phytochemicals and biomolecules such as saponins, phenolic acids, polyphenols, terpenoids, alkaloids, tannins, proteins, polyols and more. These can act as both reducing, capping and stabilizing agents during the one-step green synthesis process of ZnO nanoparticles [23]. Already many ZnO nanomaterials with various sizes and shapes have been synthesized using the extracts of different parts of the plants in recent years. Table 1 summarizes the specific sources of these plant extracts with the functional groups present therein along with different sizes and shapes of the synthesized ZnO nanoparticles depending on the type of extracts. The name written in brackets indicates the common name of the plants. The plant-based green synthesis method requires fresh leaves, fruits, seeds, stems and roots etc. which are easily available either in local markets or in gardens. After collecting them, they are thoroughly washed in fresh running water, followed by a deionized (DI) water rinse to remove any unnecessary particulate residue from these parts of the plant. After that, depending on the parts of the plant, they are cut or crushed to get small pieces. A certain amount of plant parts is taken and mixed into DI water in a beaker and heated the mixture to about 70°C to 80°C for about one hour. Finally, the solution is cooled and filtered with high quality filter paper (e.g., Whatman no. 1) and stored in a cool place (~ 4°C) as a stock solution for future use (all steps are shown in Figure 4).

Table 1. Sources of plant extracts including the functional groups contained therein involved in the production of ZnO nanoparticles of different shape and sizes

Plant extracts	Zinc precursors	Size [nm] and Shape	Functional groups	Ref.
Aloe barbadensis Miller (Aloe vera)	Zinc nitrate [$Zn(NO_3)_2$]	25-45; Spherical	–OH of intramolecular H-bond, C=O and C–C of alkanes, O–H of alcohols and C–H of alkanes, C–N of aliphatic and aromatic amines, the amide of protein and enzymes.	[24]
Parthenium hysterophorus L. (Parthenium weed)	Zinc nitrate [$Zn(NO_3)_2$]	27±5 and 84±2; Spherical and hexagonal	N-H bending and stretching modes, phosphorus compound, secondary sulfonamide, monosubstituted alkyne, amine salt.	[25]
Nephelium lappaceum (Rambutan)	Zinc nitrate hexahydrate [$Zn(NO_3)_2.6H_2O$]	Diameter ~ 450 nm, length 5 μm; Needle like	O-H stretching, H-O-H bending vibrational functional groups.	[26]
Solanum nigrum (Black nightshade)	Zinc nitrate [$Zn(NO_3)_2$]	29.79; Quasi-spherical	N–H Vibration, O–H stretching, aldehydic C–H stretching, amide III band of the random coil of protein, carboxyl side group, C-N of amine.	[27]
Vitex negundo (Chinese chaste tree)	Zinc nitrate hexahydrate [$Zn(NO_3)_2.6H_2O$]	75-80; Spherical	OH, C-H and C = C.	[28]
Allium sativum (Garlic)	Zinc nitrate hexahydrate [$Zn(NO_3)_2$. $6H_2O$]	14; Spherical	O-H bending, C-O-H bending of carboxylic group, C-N of amines, asymmetric stretching of C–H groups, C=O of polyphenols, S=O group, S-C of organosulfur compound.	[29]
Allium cepa (Onion)	Zinc nitrate hexahydrate [$Zn(NO_3)_2$. $6H_2O$]	21; Spherical	O-H bending, C-O-H bending of carboxylic group, C-N of amines, –CH_2OH groups of carbohydrates, left-handed helix of DNA (Z form).	[29]

Table 1. (Continued)

Plant extracts	Zinc precursors	Size [nm] and Shape	Functional groups	Ref.
Petroselinum crispum (Parsley)	Zinc nitrate hexahydrate $[Zn(NO_3)_2. 6H_2O]$	70; Spherical	O-H bending, C-O-H bending of carboxylic group, C-N of amines, C=O group of polyphenols and carbohydrates groups.	[29]
Azadirachta indica (Neem)	Zinc acetate dihydrate $[Zn(CH_3COO)_2].2H_2O$	9.6-25.5; Spherical	Amide II group, C–N of aliphatic, aromatic amides and aliphatic amines, alcohol and phenolic groups and stretching vibrations of secondary amine, C–H of alkanes, C–H of aromatics, C=C–H of alkynes and –OH of intramolecular H-bond, C=O and C–C of alkanes.	[30]
Corymbia citriodora (Lemon-scented gum)	Zinc nitrate hexahydrate $[Zn(NO_3)_2. 6H_2O]$	20-120; Polyhedron	O-H and N-H functional groups, C-H stretching of alkanes, C-H of aromatics, -C=C-H alkynes, C=C bending and C-C stretching of aromatic rings, C-N of aliphatic amines.	[31]
Agathosma betulina (Buchu)	Zinc nitrate hexahydrate $[Zn(NO_3)_2. 6H_2O]$	12-26; Quasi-spherical	Hydroxyl O-H functional group.	[32]
Citrus limon (Lemon)	Zinc acetate $[Zn(CH_3COO)_2]$	50; Semi-spherical	O–H group, asymmetric and symmetric vibrations of –COOH functional group, One transmission band due to C–O functional group; citric acid and sucrose compounds.	[33]
Moringa oleifera (Drumstick)	Zinc nitrate hexahydrate $[Zn(NO_3)_2. 6H_2O]$	16-31.9; Spherical	O-H, NH_2, H_3CO, HO-C=O and C-H of aromatic, N=C=S of thiamine based bioactive compounds, C=O, C=N, NH,C=C of aromatic and C-H, C=C, N-H functional groups.	[34]

Plant extracts	Zinc precursors	Size [nm] and Shape	Functional groups	Ref.
Phyllanthus niruri (Gale of the wind)	Zinc nitrate hexahydrate [$Zn(NO_3)_2$. $6H_2O$]	25.61; Quasi-spherical	Hydroxyl groups, C-H stretching, C–O and C–H vibration modes and aromatic aldehyde.	[35]
Calotropis procera	Zinc nitrate hexahydrate [$Zn(NO_3)_2$. $6H_2O$]	15-25; Spherical	C–N and N–H of amines, aromatic ring, C≡C stretching vibration group, O–H stretching, aldehydes, C–H and H–O–H functional groups.	[36]
Terminalia chebula (Myrobalan)	Zinc nitrate hexahydrate [$Zn(NO_3)_2$. $6H_2O$]	12; Spherical	Phenolic hydroxyl group, carbonyl group of phenolic acids and tannins, C=C of aromatic, C–O group of carboxylic acid and esters.	[37]
Limonia acidissima (Wood apple)	Zinc nitrate hexahydrate[$Zn(NO_3)_2$. $6H_2O$]	12-53; Spherical	Hydroxyl group, secondary amines, Si–O–Si of proteins, C–O of guaiacyl ring, C–C of aromatic, N–O of nitro compounds, C=C of vinyl alkyl ethers, C–CH_2 of chitosan or oleic acid, C–H of alkanes, O–H of phenol group.	[38]
Coffea (Coffee)	Zinc acetate dihydrate [$Zn(CH_3COO)_2$].$2H_2O$	20–46; Hexagonal	OH of the phenolic group or different carboxylic acids of coffee like citric acid, chlorogenic acid and caffeic acid, C–O group, C=C of polyphenols, –COOH group and –OH of phenols.	[39]
Carica edulis	Zinc nitrate [$Zn(NO_3)_2$]	50-55; Flower shaped	-OH group, aliphatic amine group, C-H rock, and C-H stretching groups.	[40]
Pongamia pinnata (Karanja tree)	Zinc nitrate hexahydrate [$Zn(NO_3)_2$ - $6H_2O$]	100; Spherical	-OH group, carbonyl group, C=O of non-ionic carboxylic groups, C-O-H group.	[41]

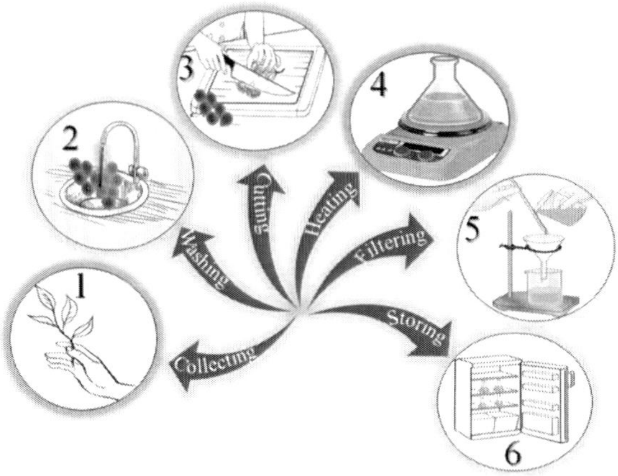

Figure 4. Key steps of plant extract preparation process.

Reaction Mechanism for the Formation of ZnO Nanomaterials

Plant extracts usually contain various phytochemicals such as polysaccharides, polyphenols, flavonoids, vitamins, amino acids, alkaloids, tannins, saponins and terpenoids etc. Most of them have antioxidant and reductive activities. For these reasons, plant extracts have been used as reducing agents, capping agents and stabilizing agents for the synthesis of nanoparticles [42, 43]. In addition, various zinc precursors, such as zinc acetate dihydrate [$Zn(CH_3COO)_2$, $2H_2O$], zinc nitrate hexahydrate [$Zn(NO_3)_2$, $6H_2O$], zinc hydroxide [$Zn(OH)_2$] and zinc chloride [$ZnCl_2$] etc. are available in the market. These precursors provide zinc ions in the solution which are reduced by these phytochemicals and precipitated as ZnO nanoparticles (Figure 5).

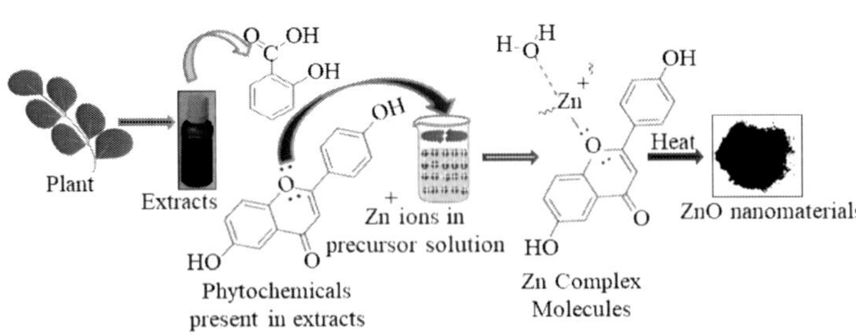

Figure 5. Proposed reaction mechanism for ZnO nanoparticle formation.

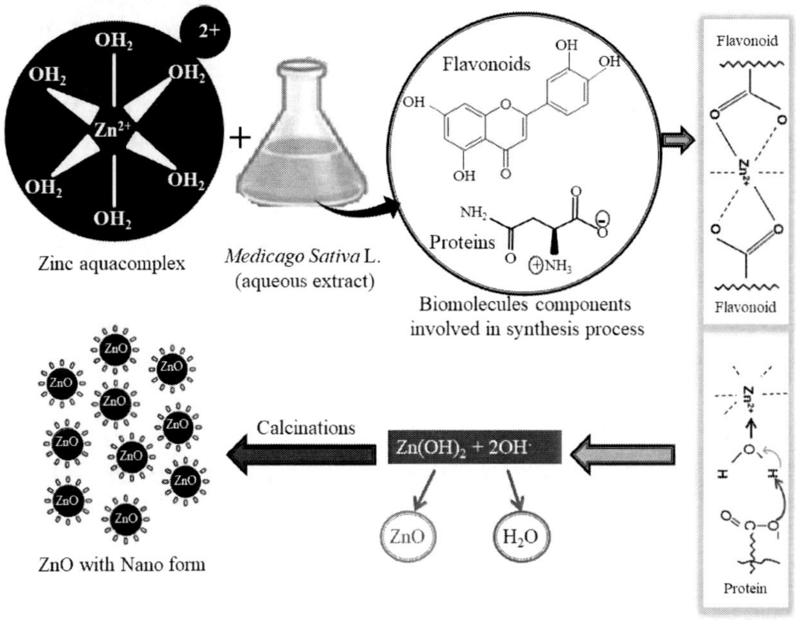

Figure 6. Schematic representation of the proposed synthesis of ZnO nanoparticles using *Medicago sativa* aqueous extracts. Adapted from Król et al. [44].

Król et al. reported a reaction mechanism for the synthesis of ZnO nanoparticles using the leaf extract of *Medicago sativa* (also known as 'Alfalfa') [44]. They used the aqueous solution of zinc nitrate hexahydrate [$Zn(NO_3)_2$, $6H_2O$] as precursor and leaf extracts as reducing agents. The mixture of the solution and the leaf extract initially formed the complex molecules after the reaction between the Zn^{2+} ions and the phytochemicals. Then, after formation of $Zn(OH)_2$ via hydrolysis, ZnO nanoparticles were obtained by calcinations (Figure 6). Matinise et al. prepared the ZnO nanoparticles using *Moringa oleifera* (also known as 'Drumstick') extract [34]. In this formation mechanism, they found that zinc nitrate hexahydrate [$Zn(NO_3)_2$, $6H_2O$] provides a large number of Zn^{2+} ions in the solution, while L-ascorbic acid, derived from *Moringa oleifera*, oxidizes to L-dehydroascorbic acid via free radicals. Then, a complex molecule of Zn-ascorbic acid was formed by electrostatic interaction between L-dehydroascorbic acid anions and Zn^{2+} cations. After calcination at high temperatures, these complex molecules turned into ZnO nanoparticles [45].

According to Osuntokun et al., when *Brassica oleracea L. var. italica* (also known as 'Broccoli') extract was used to synthesize the ZnO nanoparticles from the zinc chloride [$ZnCl_2$] solution hydroxyl functional groups on the phenolic moiety of quercetin are probably associated with Zn^{2+} to generate $Zn(OH)_2$ complex molecules. Then $Zn(OH)_2$ is converted into ZnO nanoparticles after drying in an oven heated to 70°C and subsequent calcination in a muffle furnace at 450°C [46].

Karnan et al. reported in their literature that when nanoparticles are formed using the peel extracts of *Nephelium lappaceum* (also known as 'Rambutan'), a combination of aromatic hydroxyl groups and Zn^{2+} ions in the solution results in the formation of a stable Zinc-ellagate complex molecules at pH 5–7. Zinc-ellagate complexes are then converted into ZnO nanoparticles by calcination at 450°C [47].

This phenomenon is also supported by other researchers. From their studies, it can be easily ascertained that aromatic hydroxyl groups present in the plant extracts are capable of forming complex molecules reacting with the metal ions. These complex molecules eventually break down during the calcination process, creating metal oxide nanoparticles [48].

Characteristics and Properties of Green Synthesized ZnO Nanomaterials

Not only because of various unique properties, but also due to the sustainability of our biosphere, the synthesis of ZnO nanomaterials by the plant-based green method has become the most attractive method for young researchers and scientists. To optimize the applicability of ZnO nanomaterials in the environment and on animal health, including human health, confirmation of contamination during the synthesis process, determination of size and shape with increased surface-to-volume ratio and investigation of their other unique properties are required. A series of experimental techniques can be applied to characterize and determine the different properties of synthesized ZnO nanomaterials, which can be in powder form or suspended in solution.

Compositional Analysis

The compositional analysis is the primary concern to determine the purity of the synthesized ZnO nanomaterials. This analysis helps to identify whether or not some type of contamination is present in the sample before it is used in a specific application. In this regard, many direct or indirect methods are available. For direct visualization, energy dispersive X-Ray (EDX) spectra can be used. This spectrum provides the information about the chemical compositions present in the sample [shown in Figure 7].

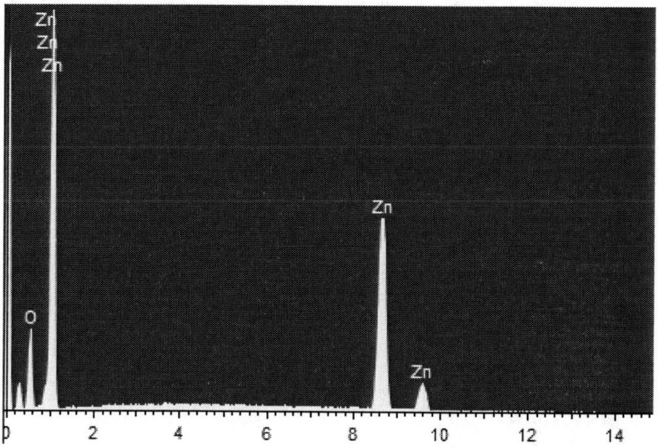

Figure 7. EDX spectra of the ZnO nanoparticles synthesized by *Azadirachta indica* leaf extracts [15].

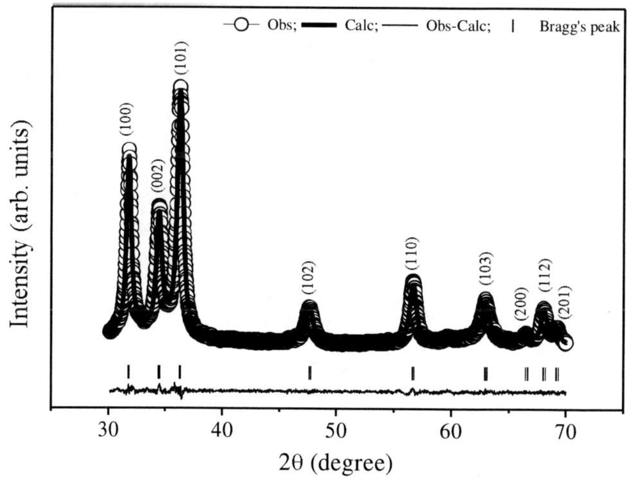

Figure 8. XRD pattern of ZnO nanostructures with the Rietveld pattern. The numbers in the brackets represent the crystal planes of the wurtzite ZnO crystal structure [15].

The oxidation state of the metal nanomaterials can also be understood by using X-ray photoelectron spectroscopy (XPS). The Fourier transform infrared spectroscopy (FTIR) is used to identify the chemical and biological functional groups in the sample. The phase identification and structural parameters of the sample could be determined by the X-ray diffraction (XRD). The obtained XRD peaks help to find the phase and other necessary information of the sample by matching the peaks using the standard JCPDS database of crystals followed by Rietveld refinement method. Figure 8 shows the Rietveld output pattern along with the observed XRD pattern, where the bubble trace represents the observed XRD pattern, the solid line through the bubble trace denotes the calculated pattern, while the zigzag line below the XRD pattern represents the difference between the observed XRD pattern and the Rietveld calculated pattern. The vertical lines above the zigzag line represent the position of the expected Bragg peaks of the wurtzite ZnO structure.

Size Determination

Size is the most important feature of nanomaterials, which makes them behave differently. There are two main factors of nano sized materials: quantum confinement and surface-to-volume ratio. Smaller particles mean more confinement, leading to an increase in the band gap, which results in a modification of the band structure of the material. Also, smaller the particles, the more percentage of surface area of individual atom or molecule are exposed to the environment, making them more chemically active. The size of the ZnO nanomaterials can be considered in terms of diameter (for spherical particles), length (for rod-shaped particles) and width (for disc-shaped particles) etc. There are many experimental techniques to determine particle size, such as Electron Microscopy (FESEM, TEM), Laser Diffraction, X-ray Diffraction (XRD) and Dynamic Light Scattering (DLS) etc. Most nanomaterials are found into a polycrystalline form containing a large number of grains commonly known as crystallites. If somehow single-crystal ZnO nanoparticles are produced, they should contain only a single grain or crystallite. For this case, the obtained particle size and crystallite size should be the same. The mean crystallite size could be determined from the XRD data using the Scherrer equation. The Williamson-Hall method also helps to determine the size of the crystallites along with the strain and stress of the crystals. Field emission scanning electron microscopy (FESEM) images show the physical representation of the particles mostly in an agglomerated form, while the Transmission electron microscopy (TEM) images are usually able to show the exact grain size and shape of the nanoparticles due to its higher resolution.

Crystallography

ZnO crystals can exist in three forms of crystal structures: cubic *rocksalt*, cubic *zinc blende*, and hexagonal *wurtzite*. Figure 9 shows the schematic diagram of these three structures. The shaded circles symbolize the constituent zinc and oxygen atoms, respectively. The large circles represent the zinc atoms while the small circles represent the oxygen atoms written inside. The *wurtzite* ZnO structure has the highest thermodynamic stability under ambient conditions, while *rocksalt* ZnO can be obtained at high pressure (~10 GPa) and the *zinc blende* ZnO structure can only be stabilized by growing on cubic substrates [49, 50]. Figure 8 represents the typical XRD pattern of a nanostructured ZnO sample. There are nine major peaks representing the ZnO crystal planes (100), (002), (101), (102), (110), (103), (200), (112) and (201) at 2θ positions 32°, 34°, 36°, 48°, 57°, 63°, 66°, 68° and 69°. After matching the peaks to the standard JCPDS database, it can be concluded that this represents the structure of hexagonal *wurtzite* ZnO with $P6_3mc$ space group.

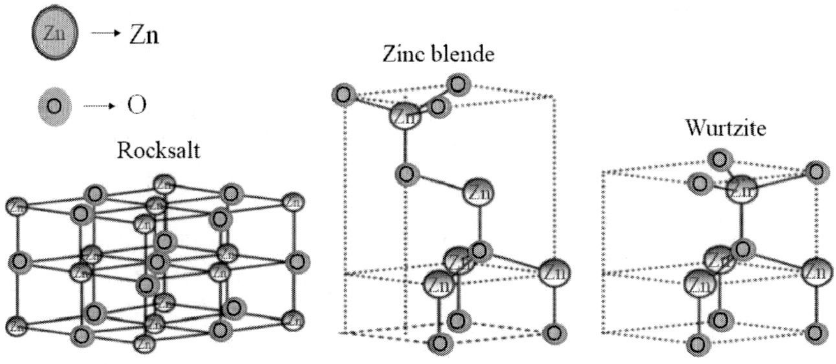

Figure 9. Three different forms of ZnO crystals: cubic rocksalt, cubic zincblende and Hexagonal-Wurtzite [51].

Surface Area of Nano Range Particles

As already discussed in section 3.2. that metal oxide nanoparticles exhibit inimitable properties due to their small size compared to their bulk form. Because of their small size, they have large surface-to-volume ratio. The surface area of an object divided by its volume is called the surface-to-volume ratio (S/V). For a very simple example considering the spherical particle of radius *r*, the surface-to-volume ratio is *3/r*. Therefore, for spherically symmetrical particles this S/V ratio is inversely proportional to its radius. When the diameter of a spherical nanoparticle is reduced from 100 nm to 1 nm, its S/V ratio increases from 6×10^{-2} nm^{-1} to 6 nm^{-1}, which is 100 times larger (represented in Figure 10). The similar phenomenon will occur for cubic particles. In addition, a spherical particle of ~100 nm has a very small percentage (~2%) of its total constituent atoms or molecules exposed to the surface. When the particle diameter is below ~10 nm, about 20 – 25% of the materials atoms are exposed to the surroundings, and when the size drops further below about 3 nm, the percentage of exposed atoms on the outer surface increases to 45 – 60%. Figure 11 shows the schematic pictorial representation to support this fact. Exposing more ZnO molecules to the surface of the nanoparticles leads to change in particle's nature in terms of chemical activity and behaves as an excellent reactor in dye degradation and an excellent sensor material for gas and chemical detection. Due to the higher S/V ratio, the melting point of nanocrystals is also lowered [52].

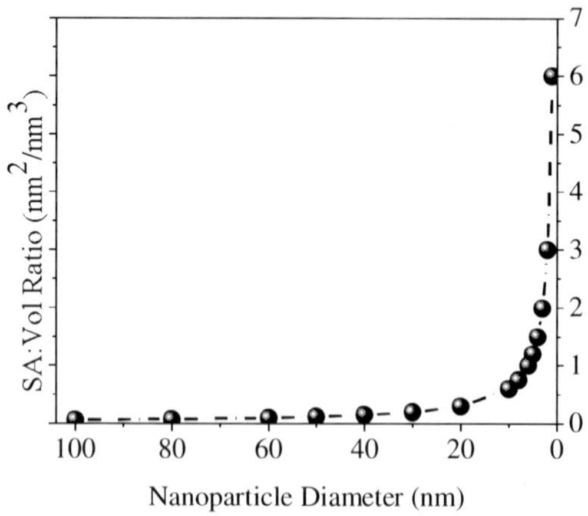

Figure 10. Variation of S/V ratio with the size of the nanoparticles.

The determination of the surface area of the nanoparticles depends on the nature of the particles: either in solid or liquid form. For the solid form, a nitrogen adsorption-desorption measurement based on Brunauer-Emmett-Teller (BET) isotherm is most commonly used. For the suspended in liquid form, simple titration and Nuclear Magnetic Resonance (NMR) techniques can be used.

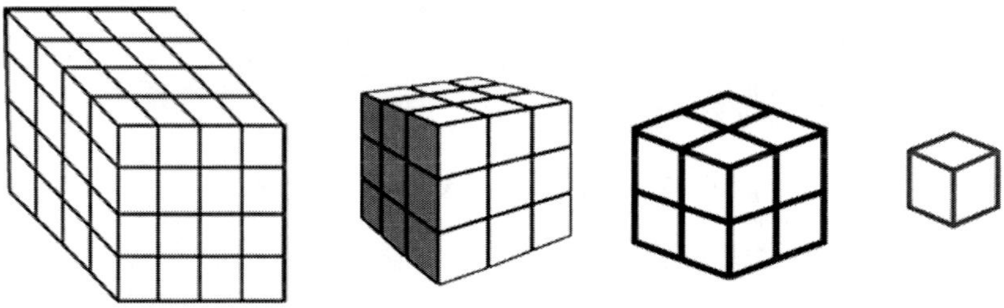

Figure 11. Schematic diagram showing the fact that more percentage of individual atoms/molecules is exposed with particle size reduction.

Surface Topography

Sometimes it is necessary to map the topography of the synthesized ZnO nanomaterials to understand their shape, size and textures. In this case, Scanning Electron Microscope (SEM), Transmission Electron Microscope (TEM) and Atomic Force Microscopy (AFM) can help to determine the surface morphology of the prepared nanomaterials. SEM is a type of electron microscope that works with a beam of electrons that interacts with the atom of the target sample and produces some signals that contain information about the surface topography. TEM is also an electron microscope, where the electron beam is passed through the sample and is capable to produce higher resolution images. AFM is able to generate a three-dimensional image of the sample surface. On the other hand, different morphologies of ZnO nanoparticles synthesized from different plant extracts have been reported in different scientific articles. Therefore, it can also be concluded that the shape and size of the nanoparticles also depend on plant extracts which we use during their synthesis. Figure 12 shows the conical shape of ZnO nanoparticles synthesized using the *Azadirachta indica* leaf extracts [53].

Figure 12. SEM image of ZnO nanoparticles synthesized with *Azadirachta indica* leaf extracts [53].

Band Gap Determination

For semiconductor materials, the band gap energy (E_g) plays an important role in determining its electrical and optical properties. Since the nano-sized metal oxides have a different band gap structure compared to their bulk form due to the effect of quantum confinement, determining the band gap of the synthesized ZnO nanomaterials is very essential before using them in various applications purposes. In this case, the UV-Visible absorption spectra can help to determine this band gap energy. The UV-Visible spectra can be obtained by passing the light through the sample suspended in a liquid and observing how strongly the substance absorbs the light. Light in the wavelength range from UV to visible (usually 190 nm to 900 nm) is used. Also, another plot called the Tauc plot is required to measure the band gap of the ZnO nanomaterials.

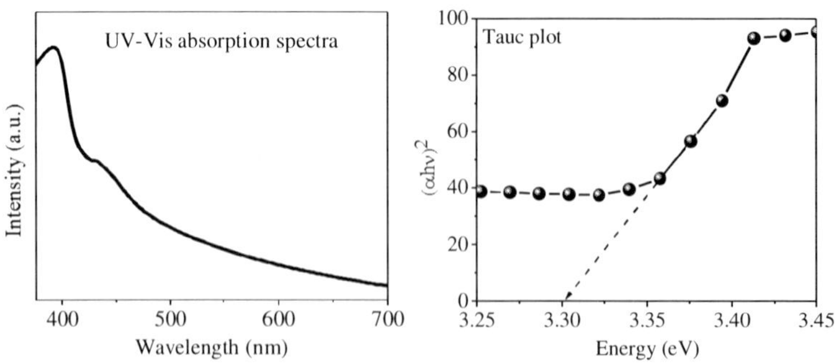

Figure 13. UV-Visible absorption spectra and Tauc plot for nano-sized ZnO [53].

For example, Figure 13 shows two graphs: (a) UV-Vis absorption spectra and (b) Tauc plot for the ZnO nanoparticles prepared using the *Azadirachta indica* leaf extract. The obtained energy band gap value is 3.30 eV [53].

Applications and Advantages of ZnO Nanomaterials

Applications of ZnO Nanomaterials Prepared by Traditional Methods

Usually, ZnO in white powder form is used as an additive in various products such as food, sunscreen, paint, lubricant, battery and sealant etc. [54, 55]. Although ZnO is mainly available in nature in the form of mineral zincite. Most of the ZnO nanoparticles, commonly required for various purposes, have been prepared in the laboratory by chemical and physical processes under different conditions. Because of its unique physical and chemical properties, including high chemical stability, high optical stability, high electrochemical coupling coefficient and a wide range of radiation absorption, ZnO nanoparticles can demonstrate its suitability for multipurpose applications shown in Figure 14 [56, 57]. Since ZnO nanomaterials exhibit strong photosensitivity, these nanomaterials can be used in light-emitting devices (LEDs) [58]. Due to its chemical sensitivity to various adsorbed gases, it can be used in gas sensing applications [59]. At room temperature, the exciton binding energy of ZnO is almost 60 meV, making it favorable for optoelectronics, field emitters, surface acoustic devices and so on [60, 61]. The use of semiconductor photocatalysis technology to eliminate groundwater pollution was first proposed by researchers in 1972 [62]. Here, with the help of light energy, semiconductor nanomaterials break down pollutants in the water. For this purpose, ZnO exhibits excellent performance. It can be used for the purpose of optoelectronic devices because of its large band gap energy (3.37 eV) [63]. ZnO and its derivatives can also be considered as additives in automotive lubricants to reduce its consumption as well as oxygen corrosion [64]. Recently, ZnO has also been used in criminological and mechanical fingerprint analysis [65]. In addition, ZnO nanomaterials are very effective and commonly applicable as crosslinking agents for carboxylated elastomers, which can contribute to the production of vulcanized rubber with high tensile strength, hardness, tear resistance and hysteresis [66]. From the view point of non-toxicity and environmental friendliness, ZnO nanomaterials are becoming more and more advanced for various application fields related to living beings. Since these materials can effectively absorb UV rays, their addition can improve the quality of cosmetics such as sunscreens and lotions [67]. The residues after applying these ZnO-based sunscreens must not cause any harm to the corals in the ocean. On the other hand, since zinc can provide important nutrients for human health, scientific research on pure ZnO is becoming more and more interesting. ZnO also has antibacterial and antifungal activities, which also allows it to be used as a food additive, food manufacturing and food packaging [68]. Recently, there have been multiple reports that demonstrate the antimicrobial activity of ZnO nanoparticles.

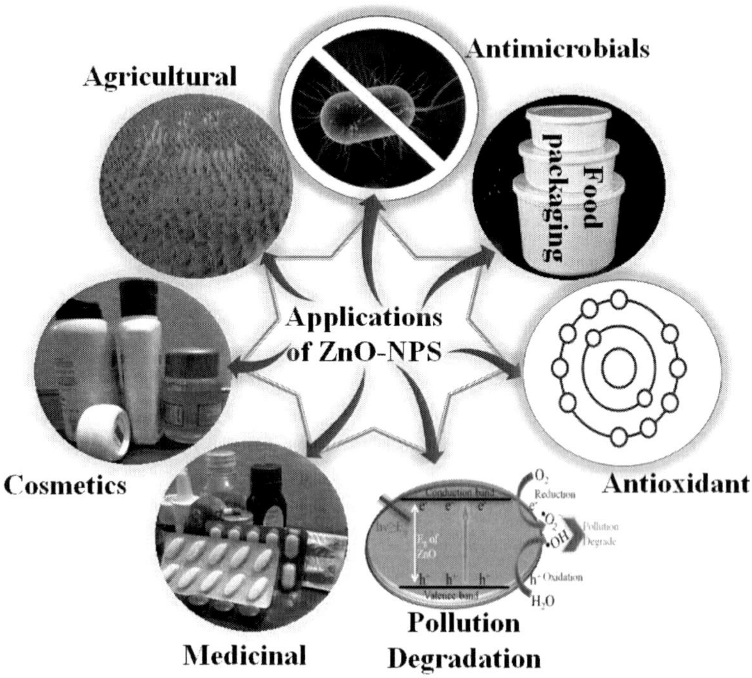

Figure 14. Different applications of ZnO nanoparticles [69].

Applications of ZnO Nanomaterials Synthesized Using Different Plant Extracts

The plant-based green synthesis of ZnO nanoparticles is also gaining importance for qualitative application in various fields such as photocatalysis, cosmetic industry, sensors, food packaging, agriculture, LEDs and solar cells etc.

In biological point of view, the use of natural plant extracts as reducing, capping and stabilizing agents during the synthesis of ZnO nanomaterials give them better biocompatibility and helps them to react with biological tissues through the interactions with phytochemical functional groups. These biological functions developed by plant extracts greatly enhance the performance of plant-based green synthesized ZnO nanomaterials. In the following subsections, the different applications of plant-based green synthesized ZnO nanomaterials are discussed in details.

ZnO Nanomaterials as Antibacterial Agent

Infections due to bacteria are a major issue to human health. Among so many metal oxide nanoparticles, ZnO nanoparticles are the most suitable candidate to be the most promising antibacterial agents because of their unique physicochemical properties. It is reported that ZnO nanoparticles are too safe and tolerable for human body. Nowadays, for antibacterial purposes, researchers are more focusing their works on the plant-based green synthesized ZnO nanoparticles compared to conventional ZnO nanoparticles because of their high purity, biocompatibility, and increased bacterial resistance to antibiotics, due to the synergistic nanostructure combining the antibacterial activity of the plant extract and the ZnO nanoparticles themselves. Several published articles describe the possible mechanism of the antibacterial action of ZnO nanoparticles. There are three possible antibacterial mechanisms: (a) The production of reactive oxygen species (ROS) such as hydroxyl radicals (OH^*) and peroxide (O_2^{-2}), which induce destruction of the cell membrane by oxidative stress and eventually damage DNA, resulting in death of the bacteria; (b) degradation of ZnO nanoparticles produces Zn^{2+} ions that interact with the bacterial cell membrane, and nucleic acid that leads to bacterial cell death; and (c) The electrostatic force between the ZnO nanoparticles and the cell membrane of the bacteria damages the plasma membrane and causes leakage of intracellular elements [70, 71]. Figure 15 shows the proposed mechanism of the antibacterial activity of the plant-based synthesized ZnO nanoparticles. There are so many articles about the ZnO nanoparticles fabricated by the plant-extract, which showed significant antibacterial activity against gram positive and gram negative bacteria such as *Staphylococcus aureus*, *Bacillus subtilis*, *Escherichia coli* and *Salmonella typhi* etc. [23, 71]. This could be due to the tiny size which allows them to produce a large number of bonds with ligands on the surface of ZnO nanoparticles and hence, with functional groups of plant extracts as well as with receptors present on the bacterial membrane, followed by invading the cytoplasm and producing multiple ROS that can permanently damage the DNA and other cellular components of the target bacteria [72-74].

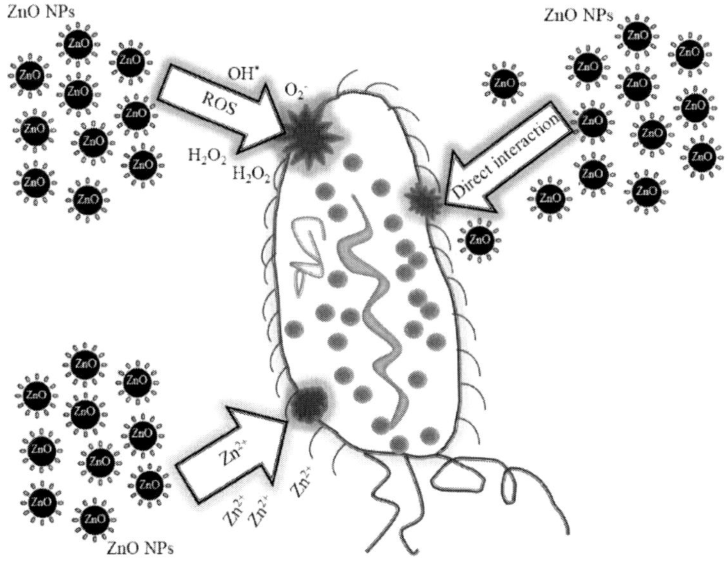

Figure 15. Proposed mechanism for the antibacterial activity of the ZnO nanoparticles.

ZnO Nanomaterials as Nano-Fertilizer

In order to meet the needs of the growing population, it is necessary to increase the yield of crop growth, which is limited by the use of traditional fertilizers. Furthermore, applications of most traditional fertilizers are not properly suited to plants. When these fertilizers are used, they can be degraded by leaching, photolysis, hydrolysis and decomposition without effective works for plant growth. Therefore, to reduce the loss of soil nutrients during fertilization and improve the crop yields, a new or alternative technology and its application based on nanoscale size are needed [23]. On the other hand, since Zinc (Zn) is an essential element for the healthy growth of plants, animals and humans, its deficiency can cause many issues [75]. Zn is mainly required by plants for the purpose of carbohydrate metabolism as well as gene expression in retort to environmental stress. Usually, soluble Zn salts are used as fertilizers. This type of fertilizer assists in the assimilation of Zn, which can cause many environmental issues [76]. According to the researchers, plant-based ZnO nanomaterials with biomolecular ligands could be used as novel growth promoters to support high crops productivity. Zn is a metal element that can be present in six types of enzymes such as oxidoreductase, transferase, lyase, isomerase, hydrolase and ligase. The carotenoid, "chlorophyll a" levels, total amount of plant biomass, can be enhanced by changing the concentrations of ZnO nanomaterials. These improvements promote the biosynthesis of chlorophyll, carotenoids and ultimately plant growth. Plant-based green synthesized ZnO nanomaterials can be used as nano-fertilizers to provide micronutrients, since ZnO nanoparticles are unable to provide the essential micronutrient even without co-factors [77, 78]. In this nano-fertilization process, the ZnO nanomaterials simply dissolve in the liquid and release the nutrients in the form of ions. Porous transport process transports these nutrients to the desired location. Compared to conventional fertilizers, green

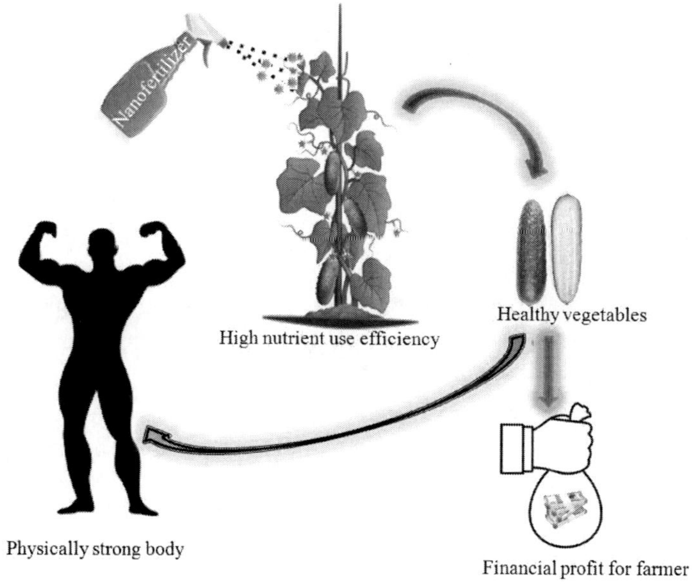

Figure 16. Benefits of nano-fertilizers.

synthesis ZnO-based nano-fertilizers achieve better plant growth, improve fertilizer quality for a large number of crops, reduce nutrient loss and trim down the probability of penetration not only into the surface but also into the groundwater and reduce the risk for human health. Furthermore, green synthesized ZnO-based nano-fertilizers give us another chance for greener sustainable development, which can establish harmonious connection between development and ecology. As green synthesized ZnO-based nano-fertilizers exhibit slow-release behavior, they show efficient work with small amounts, which can help to reduce fertilizer consumption with environmental safety [79]. Figure 16 describes the benefits of ZnO nano fertilizers prepared by plant extracts.

Photocatalytic Activity of ZnO Nanoparticles

ZnO nanomaterials prepared using the plant extracts has an inherent ability to absorb UV and other solar radiations. For this reason, they can exhibit exceptional degradation behavior of so many dyes such as methyl orange, methylene blue etc. Textile industries across the country discharge so many toxic elements and runoff into rivers or lakes near the factory, causing harm to marine life, which ultimately harms the ecosystem. This is nothing but a serious environmental crisis. The mechanism for photodegradation is that when ZnO nanomaterials, which are also a semiconductor, absorbs energy (UV light or sunlight) greater than their band gap energy, electron jumps from valence band to conduction band cause charge separation into the catalyst, leads to the formation of holes in the valence band and electrons in the conduction band [80-83]. The generated electron-hole pairs spread on the surface of the ZnO photocatalyst, allowing it to participate in the chemical reaction. The oxidizing effect between the free electron-holes and the surrounding oxygen or water molecules creates highly reactive hydroxyl (OH^*) radicals that aid in the degradation of the contaminated water's colorants [84]. Figure 17 represents the proposed mechanism for the photocatalytic dye degradation by the ZnO nanomaterials.

Several articles have been published on the photocatalytic activity of plant-mediated synthesized ZnO nanoparticles as a catalyst in the presence of solar light. Alamdari et al. observed the significant 80% degradation of the methylene blue dye after 200 min of UV irradiation [85]. Here they used ZnO nanoparticles which were synthesized using *Sambucus ebulus* leaf extracts. Vinayagam et al. also reported the successful degradation of the rhodamine B dye by up to 94.74% within 165 minutes after sun exposure by ZnO nanoparticles prepared using the *Cyanometra ramiflora* leaf extracts [86]. The specific properties like mesoporous crystalline nature, large surface area and enormous electron-generating capabilities of ZnO nanoparticles with the presence of functional groups from plant extracts probably support the rapid and accelerated dye degradations.

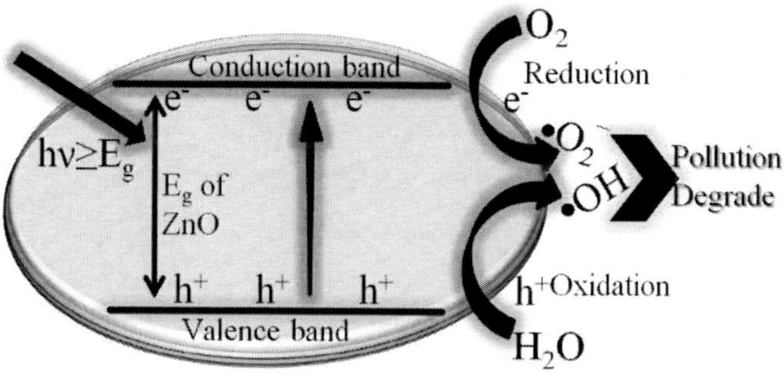

Figure 17. Mechanism for the photocatalytic dye degradation by the ZnO nanoparticles.

Drug Delivery Activity of ZnO Nanoparticles

Drug delivery is a method of transporting drugs or any pharmaceutical compounds into the human or animal body to produce a therapeutic effect. Drugs can be delivered to the body in numerous ways – by swallowing, inhaling, absorbing through the skin, or by injection. Each method has their own pros and cons. Depending on the type of disease, different drugs are administered to the body. Recent research has revealed that ZnO is a potential candidate for drug delivery due to its non-toxicity, biocompatibility, easy cellular internalization and porous structure. To attack the malignant cells in the animal body, the metal oxide nanoparticles are the best option because they are very small in size (1 – 100 nm) with larger S/V ratio which makes it easier to penetrate cell membranes and be absorbed by the cells, resulting in better distribution into the body. They can be easily transmitted through the blood-brain-barrier (BBB), enter the pulmonary system, and be taken up through the junctions of the skin's endothelial cells [87]. Besides of these, ZnO nanoparticles release Zn^{2+} ions and produce reactive oxygen species (ROS) that cause damage to proteins, DNA and lipids. As a result, the release of the targeted drugs acts on specific cells and causes cell damage or even death [88]. Figure 18 shows the proposed mechanism for the targeted drug delivery method of the plant-based green synthesized ZnO nanoparticles. Here nanoscale ZnO can produce a large number of bonds with ligands on the surface of nanoparticles and hence, with functional groups of plant extracts as well as with receptors present on the cell membrane and thus help to drug delivery.

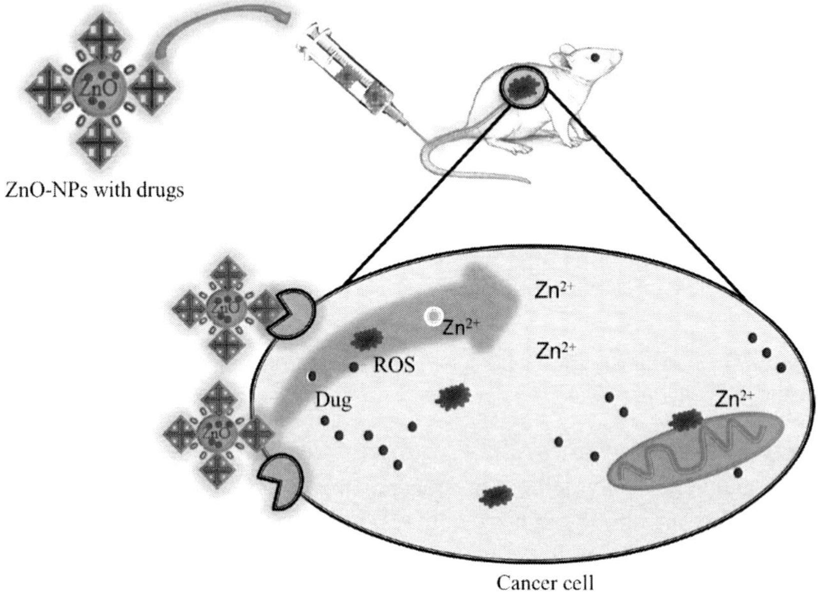

Figure 18. Plausible mechanism for the targeted drug release method of the plant-based ZnO nanomaterials.

Vimala et al. reported in their article that they have successfully synthesized ZnO nanoparticles using *Borassus flabellifer* fruit extract. After successfully uploading doxorubicin onto ZnO nanoparticles and evaluated their biological activity successfully in killing cancer cells of leukemia [89]. Another report of Akbarian et al. [90] shows a successful loading of an essential anticancer drug onto chitosan-coated ZnO nanoparticles prepared from the *Camellia sinensis* leaf extract and showed successful biological activity as a drug delivery system on breast cancer cell lines.

Despite some reports on the use of ZnO nanoparticles in targeted drug delivery systems, further research is needed to develop plant-based ZnO nanoparticles as a flexible drug for targeted delivery.

Advantages of Plant Extract-Based Green Synthesized ZnO Nanomaterials

Numerous articles have been published claiming that the plant-based synthesis of ZnO nanomaterials have so many benefits. Plants are the actual source of phytochemicals such as polyphenols, flavonoids, quercetin, etc., which are non-hazardous and take part in the synthesis process, replacing the harmful chemicals used in the other traditional methods. In addition, the waste products after synthesis are biodegradable and environment-friendly. This synthesis process is a simple, ambient conditioned, one-step process, time saving, and uses simple and inexpensive equipment. Also, in this synthesis methods there are so many possibilities like reaction time, amount and density of plant extracts and precursor solutions, pH value, temperature, pressure, calcination temperature, time etc., to control the shape and size of the nanoparticle to be produced. This could undoubtedly be used as a large-scale production process in the near future. The ready availability of plant-extracts, easy to available precursors, one-step manufacturing technique, time-saving process and inexpensive equipment for synthesis help this method to meet the socio-economic and environmental requirements. The tremendous potency of plant-based synthesis of ZnO nanomaterials can drive the progress of joint development of society, environment and scientific research.

Future Prospects and Challenges

The foregoing discussions substantiate that the plant-based green synthesis method is a new production method for fabricating ZnO nanomaterials after the traditional methods for preparing nanoparticles. This plant-based method also has the ability to produce nanoparticles with different morphology, shape and size along with various applications ranging from animal health to cosmetics, agriculture and food etc. This plant-based green synthesis of ZnO nanomaterial is very suitable to promote sustainable development based on nanoscience and nanotechnology.

In the introductory part, the statistics of the number of published scientific articles of the last twenty-two years (2000 to 2022) shows the growing interest of the new researchers in crafting the plant-based ZnO nanomaterials and to apply them into nanomedicine, nanoscience and nanotechnology. The upward trend clearly helps to predict that this interest will grow much more in the near future and last for a long time for reasons such as not only using harmless raw materials but also producing biodegradable and less hazardous by-products with relatively inexpensive equipment compared to other traditional methods.

Despite all these benefits, there are a few limitations and shortcomings that need to be solved in the near future.

i. *Easy availability of plants*

Most researchers have studied the plant-based synthesis of nanoscale ZnO particles using the plant extracts, which are mainly locally available. These studies reported the use of plants mainly localized in specific counties or regions due to the compatibility of soil and environmental conditions. It seems to be difficult to establish the large-scale global production of particular plant-based green synthesis of ZnO nanoparticles. Therefore, the possibility to use the local evergreen plants only in large-scale production of ZnO nanomaterials is yet to be explored.

ii. *Time constraints*

Most of the plants and parts of plants can be collected in a specific season; their growing time is very different in various countries/ regions or at different altitudes. The flowers for synthesizing ZnO nanomaterials can only be collected during a specific flowering period. On the other hand, seeds of a specific plant can only be collected during fruiting season. Therefore, these types of issues limit the availability of different parts of the plant. Evergreen plants like *Azadirachta indica*, *Citrus limon*, *Moringa oleifera* etc. could avoid the seasonal time constraints. Additionally, raw materials that are not limited by seasonal and geographical availability should be the primary research focus for further exploration. When selecting the required raw materials, the researcher should first consider the geographical and seasonal abundance for material sourcing.

iii. *Choosing the right phytochemical*

It is reported that the phytochemicals play the key role in the production of ZnO nanoparticles. However, the shape and size of the ZnO nanoparticles synthesized with different plant-extracts are very different. On the other hand, the information about the actual activity of the different phytochemicals is still insufficient. There are so many reports on the synthesis of ZnO nanoparticles using the extract of a specific plant part, with the particle size and morphology having different values. It is, therefore, necessary to know which plant extracts are to be used and in what quantity for the production of nanoparticles of a certain size and shape. The exact name of the phytochemicals contained in a given extract and the exact key role of the phytochemicals during the nanoparticle synthesis process is still questionable and requires extensive research. It will help in the decision to select the right phytochemicals for the production of ZnO nanoparticles of a certain shape and size.

iv. *Reaction Mechanism and Plant Extract to Precursor Salt ratio*

The exact reaction mechanism and the optimal stoichiometric ratio of certain plant extracts and the precursor salt in terms of shape and size control should be systematically investigated.

v. *Synthesis factors*

Future research should focus on the synthesis factors such as solution pH, temperature and pressure that need to be maintained during the synthesis. These factors may have direct relation with the shape, size and morphology of the ZnO nanoparticles.

vi. *Stability and reusability of nanoparticles*

Product stability is also a major concern in nanoparticle quality. It is necessary to find the finest method to obtain the ZnO nanomaterials from their synthesis to their preservation, in order to make them more stable for a long time with the abilities to prevent moisture absorption, without any sign of sedimentation and ultimately without losing their activity and still remain biocompatible, ever for a longer time (e.g., a year or two). The reusability of the green-based ZnO nanomaterials must also be considered. Further investigations are being considered to make the ZnO nanomaterials so that they can be reused over and over again (at least 3-5 times) without losing their activity. Also, the storage and the preservation of the ZnO nanoparticles should be carefully considered. If the nanoscale ZnO materials can be kept under ambient conditions, the cost of storage and preservation will be greatly reduced.

vii. *Commercialization*

ZnO nanomaterials based on plant extracts yet not adequately and widely commercialized in various fields such as drug delivery, food packaging and cosmetics. Research and development section should concentrate their attention on this purpose such that green-based ZnO nanoparticles can be commercialized for greener future.

Conclusion

This chapter attempts to cover in detail almost all the important greener aspects of the plant-based green synthesis of ZnO nanomaterials including their greener applications. The above discussion proves that the plant-based green synthesis method is suitable for an alternate and large-scale production technique with higher efficiency. Since environmental pollution is increasing so fast, this plant-based nanoparticle synthesis became necessary very quickly and also promotes the development of the green revolutionary process to produce other promising nanomaterials besides zinc oxide nanomaterials. Being a plant-based method, it can have various merits such as clean, non-hazardous, reliable, biodegradable and eco-friendly approaches. The scientific review systematically reveals that the synthesis of ZnO nanoparticles using plant-extracts is practicable. However, this synthesis method also has so many challenges and shortcomings, such as low-yield, non-uniform particle sizes, seasonal and regional availability of raw materials and other difficulties that need to be overcome for the production and practical applications of plant-based synthesized nanomaterials. Therefore, improving the production of nanoscale ZnO particles, using easily available evergreen plant raw materials, and adopting simple energy-saving technologies are the research directions of the future. Hence, the green synthesis of ZnO nanomaterials has a broad perspective and great potential for sustainable development.

References

[1] Raulin, J. 1869. "Études cliniques sur la vegetation." *Ann. Sci. Nat. Bot. Biol. Veg. Ser.* 5 (B): 93.

[2] Mendel, L. B., and Bradley, H. C. 1905. "Experimental studies on the physiology of the molluscs—second paper." *American Journal of Physiology-Legacy Content*, 14(4): 313-327.

[3] Sommer, A. L., and Lipman, C. B. 1926. "Evidence on the indispensable nature of zinc and boron for higher green plants." *Plant physiology* 1(3): 231.

[4] Todd, W. R., Elvehjem, C. A., and Hart, E. B. 1933. "Zinc in the nutrition of the rat." *American Journal of Physiology-Legacy Content* 107(1): 146-156.

[5] O'dell, B. L., Newberne, P. M., and Savage, J. E. 1958. "Significance of dietary zinc for the growing chicken." *The Journal of Nutrition* 65(4): 503-523.

[6] Prasad, A. S., Miale Jr, A., Farid, Z., Sandstead, H. H., and Schulert, A. R. 1963. "Zinc metabolism in patients with the syndrome of iron deficiency anemia, hepatosplenomegaly, dwarfism, and hypogonadism." *Journal of Laboratory and Clinical Medicine* 61(4): 537-49.

[7] Siddiqi, K. S., and Husen, A. 2018. "Properties of zinc oxide nanoparticles and their activity against microbes." *Nanoscale research letters* 13(1): 1-13.

[8] Raha, S., and Ahmaruzzaman, M. 2022. "ZnO nanostructured materials and their potential applications: progress, challenges and perspectives." *Nanoscale Advances* 4(8): 1868-1925.

[9] Basnet, P., Chanu, T. I., Samanta, D., and Chatterjee, S. 2018. "A review on bio-synthesized zinc oxide nanoparticles using plant extracts as reductants and stabilizing agents." *Journal of Photochemistry and Photobiology B: Biology* 183: 201-221.

[10] Rochman, N. T., and Akwalia, P. R. 2017. "Fabrication and characterization of Zinc Oxide (ZnO) nanoparticle by sol-gel method." *Journal of physics: conference series – IOP Publishing* 853(1).

[11] Baruwati, B., Kumar, D. K., and Manorama, S. V. 2006. "Hydrothermal synthesis of highly crystalline ZnO nanoparticles: a competitive sensor for LPG and EtOH." *Sensors and Actuators B: Chemical* 119(2): 676-682.

[12] Aghababazadeh, R., Mazinani, B., Mirhabibi, A., and Tamizifar, M. 2006. "ZnO nanoparticles synthesised by mechanochemical processing."*Journal of Physics: Conference Series - IOP Publishing* 26(1): 075.

[13] Qu, J., Yuan, X., Wang, X., and Shao, P. 2011. "Zinc accumulation and synthesis of ZnO nanoparticles using *Physalis alkekengi L.*" *Environmental pollution* 159(7): 1783-1788.

[14] Alamdari, S., Sasani Ghamsari, M., Lee, C., Han, W., Park, H. H., Tafreshi, M.J., Afarideh, H. and Ara, M. H. M. 2020. "Preparation and characterization of zinc oxide nanoparticles using leaf extract of Sambucus ebulus." *Applied Sciences* 10(10): 3620.

[15] Barman, S., Sikdar, S., Biswas, A., Mandal, B. K., and Das, R. 2020. "Structural microanalysis of green synthesized $Al_xZn_{(1-x)}O$ nanoparticles." *Nano Express* 1(2): 020003.

[16] Huang, Y., Haw, C. Y., Zheng, Z., Kang, J., Zheng, J. C., and Wang, H. Q. 2021. "Biosynthesis of zinc oxide nanomaterials from plant extracts and future green prospects: a topical review." *Advanced Sustainable Systems* 5(6): 2000266.

[17] Liu, B., and Zeng, H. C. 2003. "Hydrothermal synthesis of ZnO nanorods in the diameter regime of 50 nm." *Journal of the American Chemical Society* 125(15): 4430-4431.

[18] Nam, G., Kim, M. S., Lee, J., Leem, J. Y., Lee, S. H., Jung, J. H., ... and Kim, J. S. 2013. "Growth and optical properties of sol-gel ZnO thin films grown on R-plane sapphire substrates." *Journal of the Korean Physical Society* 62(8): 1154-1159.

[19] Xia, T., Kovochich, M., Liong, M., Madler, L., Gilbert, B., Shi, H., Yeh, J. I., Zink, J. I. and Nel, A. E. 2008. "Comparison of the mechanism of toxicity of zinc oxide and cerium oxide nanoparticles based on dissolution and oxidative stress properties." *ACS nano* 2(10): 2121-2134.

[20] He, C., Sasaki, T., Shimizu, Y., and Koshizaki, N. 2008. "Synthesis of ZnO nanoparticles using nanosecond pulsed laser ablation in aqueous media and their self-assembly towards spindle-like ZnO aggregates." *Applied Surface Science* 254(7): 2196-2202.

[21] Ip, K., Heo, Y. W., Baik, K. H., Norton, D. P., Pearton, S. J., and Ren, F. 2004. "Carrier concentration dependence of Ti/Al/Pt/Au contact resistance on n-type ZnO." *Applied Physics Letters* 84(4): 544-546.

[22] Pardeshi, S. K., and Patil, A. B. 2009. "Effect of morphology and crystallite size on solar photocatalytic activity of zinc oxide synthesized by solution free

mechanochemical method." *Journal of Molecular Catalysis A: Chemical* 308(1-2): 32-40.

[23] Singh, A., Singh, N. Á., Afzal, S., Singh, T., and Hussain, I. 2018. "Zinc oxide nanoparticles: a review of their biological synthesis, antimicrobial activity, uptake, translocation and biotransformation in plants." *Journal of materials science* 53(1): 185-201.

[24] Sangeetha, G., Rajeshwari, S., and Venckatesh, R. 2011. "Green synthesis of zinc oxide nanoparticles by aloe barbadensis miller leaf extract: Structure and optical properties." *Materials Research Bulletin* 46(12): 2560-2566.

[25] Rajiv, P., Rajeshwari, S., and Venckatesh, R. 2013. "Bio-Fabrication of zinc oxide nanoparticles using leaf extract of *Parthenium hysterophorus L.* and its size-dependent antifungal activity against plant fungal pathogens." *Spectrochimica Acta Part A: Molecular and Biomolecular Spectroscopy* 112: 384-387.

[26] Yuvakkumar, R., Suresh, J., Nathanael, A. J., Sundrarajan, M., and Hong, S. I. 2014. "Novel green synthetic strategy to prepare ZnO nanocrystals using rambutan (*Nephelium lappaceum L.*) peel extract and its antibacterial applications." *Materials Science and Engineering: C* 41: 17-27.

[27] Ramesh, M., Anbuvannan, M., and Viruthagiri, G. J. S. A. P. A. M. 2015. "Green synthesis of ZnO nanoparticles using *Solanum nigrum* leaf extract and their antibacterial activity." *Spectrochimica Acta Part A: Molecular and Biomolecular Spectroscopy* 136: 864-870.

[28] Ambika, S., and Sundrarajan, M. 2015. "Antibacterial behaviour of *Vitex negundo* extract assisted ZnO nanoparticles against pathogenic bacteria." *Journal of Photochemistry and photobiology B: Biology* 146: 52-57.

[29] Stan, M., Popa, A., Toloman, D., Dehelean, A., Lung, I., and Katona, G. 2015. "Enhanced photocatalytic degradation properties of zinc oxide nanoparticles synthesized by using plant extracts." *Materials Science in Semiconductor Processing* 39: 23-29.

[30] Bhuyan, T., Mishra, K., Khanuja, M., Prasad, R., and Varma, A. 2015. "Biosynthesis of zinc oxide nanoparticles from *Azadirachta indica* for antibacterial and photocatalytic applications." *Materials Science in Semiconductor Processing* 32: 55-61.

[31] Zheng, Y., Fu, L., Han, F., Wang, A., Cai, W., Yu, J., ... and Peng, F. 2015. "Green biosynthesis and characterization of zinc oxide nanoparticles using *Corymbia citriodora* leaf extract and their photocatalytic activity." *Green Chemistry Letters and Reviews* 8(2): 59-63.

[32] Thema, F. T., Manikandan, E., Dhlamini, M. S., and Maaza, M. J. M. L. 2015. "Green synthesis of ZnO nanoparticles via *Agathosma betulina* natural extract." *Materials Letters* 161: 124-127.

[33] Davar, F., Majedi, A., and Mirzaei, A. 2015. "Green synthesis of ZnO nanoparticles and its application in the degradation of some dyes." *Journal of the American Ceramic Society* 98(6): 1739-1746.

[34] Matinise, N., Fuku, X. G., Kaviyarasu, K., Mayedwa, N., and Maaza, M. 2017. "ZnO nanoparticles via *Moringa oleifera* green synthesis: Physical properties and mechanism of formation." *Applied Surface Science* 406: 339-347.

[35] Anbuvannan, M., Ramesh, M., Viruthagiri, G., Shanmugam, N., and Kannadasan, N. 2015. "Synthesis, characterization and photocatalytic activity of ZnO nanoparticles prepared by biological method." *Spectrochimica Acta Part A: Molecular and Biomolecular Spectroscopy* 143: 304-308.

[36] Gawade, V. V., Gavade, N. L., Shinde, H. M., Babar, S. B., Kadam, A. N., and Garadkar, K. M. 2017. "Green synthesis of ZnO nanoparticles by using *Calotropis procera* leaves for the photodegradation of methyl orange." *Journal of Materials Science: Materials in Electronics* 28(18): 14033-14039.

[37] Rana, N., Chand, S., and Gathania, A. K. 2016. "Green synthesis of zinc oxide nano-sized spherical particles using *Terminalia chebula* fruits extract for their photocatalytic applications." *International Nano Letters* 6(2): 91-98.

[38] Patil, B. N., and Taranath, T. C. 2016. "*Limonia acidissima L.* leaf mediated synthesis of zinc oxide nanoparticles: A potent tool against Mycobacterium tuberculosis." *International journal of mycobacteriology* 5(2): 197-204.

[39] Koupaei, M. H., Shareghi, B., Saboury, A. A., Davar, F., Semnani, A., and Evini, M. 2016. "Green synthesis of zinc oxide nanoparticles and their effect on the stability and activity of proteinase K." *RSC advances* 6(48): 42313-42323.

[40] Fowsiya, J., Madhumitha, G., Al-Dhabi, N. A., and Arasu, M. V. 2016. "Photocatalytic degradation of Congo red using Carissa edulis extract capped zinc oxide nanoparticles." *Journal of Photochemistry and Photobiology B: Biology* 162: 395-401.

[41] Sundrarajan, M., Ambika, S., and Bharathi, K. 2015. "Plant-extract mediated synthesis of ZnO nanoparticles using *Pongamia pinnata* and their activity against pathogenic bacteria." *Advanced powder technology* 26(5): 1294-1299.

[42] Liu, Y. J., Li, J. F., and Chen, Z. L. 2019. "Research progress of green synthesis of iron nanoparticle and its application on contaminants removal from water." *Technol. Water Treat* 45: 6-11.

[43] Agarwal, H., Kumar, S. V., and Rajeshkumar, S. 2017. "A review on green synthesis of zinc oxide nanoparticles – An eco-friendly approach." *Resource-Efficient Technologies* 3(4): 406-413.

[44] Król, A., Railean-Plugaru, V., Pomastowski, P., and Buszewski, B. 2019. "Phytochemical investigation of *Medicago sativa L.* extract and its potential as a safe source for the synthesis of ZnO nanoparticles: The proposed mechanism of formation and antimicrobial activity." *Phytochemistry Letters* 31: 170-180.

[45] Matinise, N., Kaviyarasu, K., Mongwaketsi, N., Khamlich, S., Kotsedi, L., Mayedwa, N., and Maaza, M. 2018. "Green synthesis of novel zinc iron oxide ($ZnFe_2O_4$) nanocomposite via *Moringa Oleifera* natural extract for electrochemical applications." *Applied Surface Science* 446: 66-73.

[46] Osuntokun, J., Onwudiwe, D. C., and Ebenso, E. E. 2019. "Green synthesis of ZnO nanoparticles using aqueous *Brassica oleracea L. var. italica* and the photocatalytic activity." *Green chemistry letters and reviews* 12(4): 444-457.

[47] Karnan, T., and Selvakumar, S. A. S. 2016. "Biosynthesis of ZnO nanoparticles using rambutan (*Nephelium lappaceum L.*) peel extract and their photocatalytic activity on methyl orange dye." *Journal of molecular Structure* 1125: 358-365.

[48] Mayedwa, N., Mongwaketsi, N., Khamlich, S., Kaviyarasu, K., Matinise, N., and Maaza, M. 2018. "Green synthesis of nickel oxide, palladium and palladium oxide synthesized via *Aspalathus linearis* natural extracts: physical properties & mechanism of formation." *Applied Surface Science* 446:266-272.

[49] Meng, X. Q., Shen, D. Z., Zhang, J. Y., Zhao, D. X., Lu, Y. M., Dong, L., Zhang, Z. Z. et al. 2005. "The structural and optical properties of ZnO nanorod arrays." *Solid State Communications* 135 (3): 179-182.

[50] Hadis, M., and Ümit, Ö. 2009. "*General Properties of ZnO*." Wiley Online Library.

[51] Shaba, E. Y., Jacob, J. O., Tijani, J. O., and Suleiman, M. A. T. 2021. "A critical review of synthesis parameters affecting the properties of zinc oxide nanoparticle and its application in wastewater treatment." *Applied Water Science* 11(2): 1-41.

[52] Sengupta, A., and Sarkar, C. K. (Eds.). 2015. "*Introduction to nano: basics to nanoscience and nanotechnology.*" Springer.

[53] Barman, S., Kumar, J., Das, A. K., Sikdar, S., Biswas, A., Srinivasan, A., and Das, R. 2022. "Comparative Study of ZnO Nanomaterials Synthesized by Green and Electrospinning Methods." *Journal of Nano Research* 72: 81-93.

[54] Chandramohan K., Valli R., and Mageswari, B. 2017. "Synthesis and characterization of zinc nanopartical from *Lufa acutangular*." *International Journal of Science and Research* 6(11): 339–340.

[55] Perveen, R., Shujaat, S., Qureshi, Z., Nawaz, S., Khan, M. I., and Iqbal, M. 2020. "Green versus sol-gel synthesis of ZnO nanoparticles and antimicrobial activity evaluation against panel of pathogens." *Journal of Materials Research and Technology* 9(4): 7817-7827.

[56] Lou, X., Shen, H. S., and Shen, Y. S. 1991. "Development of ZnO series ceramic semiconductor gas sensors." *J. Sens. Trans. Technol* 3(1): 1-5.

[57] Segets, D., Gradl, J., Taylor, R. K., Vassilev, V., and Peukert, W. 2009. "Analysis of optical absorbance spectra for the determination of ZnO nanoparticle size distribution, solubility, and surface energy." *ACS nano* 3(7): 1703-1710.

[58] Janotti, A., and Van de Walle, C. G. 2009. "Fundamentals of zinc oxide as a semiconductor." *Reports on progress in physics* 72(12): 126501.

[59] Gouthama, S., Kaur, S., Sadasivuni, K. K., Bal, J. K., Jayarambabu, N., Kumar, D. S., and Rao, K. V. 2017. "Nanostructured ZnO gas sensors obtained by green method and combustion technique. " *Materials Science in Semiconductor Processing* 57: 110-115.

[60] Gorla, C. R., Emanetoglu, N. W., Liang, S., Mayo, W. E., Lu, Y., Wraback, M., and Shen, H. 1999. "Structural, optical, and surface acoustic wave properties of epitaxial ZnO films grown on (0112) sapphire by metalorganic chemical vapor deposition." *Journal of Applied Physics* 85(5): 2595-2602.

[61] Zhang, H., Yang, D., Ma, X., and Que, D. 2005. "Synthesis and field emission characteristics of bilayered ZnO nanorod array prepared by chemical reaction." *The Journal of Physical Chemistry* B, 109(36): 17055-17059.

[62] Qi, K., Cheng, B., Yu, J., and Ho, W. 2017. "Review on the improvement of the photocatalytic and antibacterial activities of ZnO." *Journal of Alloys and Compounds* 727: 792-820.

[63] Dhanemozhi, A. C., Rajeswari, V., and Sathyajothi, S. 2017. "Green synthesis of zinc oxide nanoparticle using green tea leaf extract for supercapacitor application." *Materials Today: Proceedings* 4(2): 660-667.

[64] Klingshirn, C. 2007. "ZnO: From basics towards applications." *Physica Status Solidi* (b), 244(9): 3027-3073.

[65] Kołodziejczak-Radzimska, A., and Jesionowski, T. 2014. "Zinc oxide—from synthesis to application: a review." *Materials* 7(4): 2833-2881.

[66] Ibarra, L., Marcos-Fernandez, A., and Alzorriz, M. 2002. "Mechanistic approach to the curing of carboxylated nitrile rubber (XNBR) by zinc peroxide/zinc oxide." *Polymer* 43(5): 1649-1655.

[67] Ju-Nam, Y., and Lead, J. R. 2008. "Manufactured nanoparticles: an overview of their chemistry, interactions and potential environmental implications." *Science of the total environment* 400(1-3): 396-414.

[68] Espitia, P. J. P., Soares, N. D. F. F., Coimbra, J. S. D. R., de Andrade, N. J., Cruz, R. S., and Medeiros, E. A. A. 2012. "Zinc oxide nanoparticles: synthesis, antimicrobial activity and food packaging applications." *Food and bioprocess technology* 5(5): 1447-1464.

[69] Zeghoud, S., Hemmami, H., Seghir, B. B., Amor, I. B., Kouadri, I., Rebiai, A., ... and Simal-Gandara, J. 2022. "A review on biogenic green synthesis of ZnO nanoparticles by plant biomass and their applications." *Materials Today Communications* 104747.

[70] Elumalai, K., and Velmurugan, S. 2015. "Green synthesis, characterization and antimicrobial activities of zinc oxide nanoparticles from the leaf extract of *Azadirachta indica* (L.)." *Applied Surface Science* 345: 329-336.

[71] Sharmila, G., Muthukumaran, C., Sandiya, K., Santhiya, S., Sakthi, P. R., Manoj, K. N., Suriyanarayanan, N., and Thirumarimurugan, M. 2018. "Biosynthesis, characterization, and antibacterial activity of zinc oxide nanoparticles derived from *Bauhinia tomentosa* leaf extract." *Journal of Nanostructure in chemistry* 8: 293-299.

[72] Ahmed, S., Chaudhry, S. A., and Ikram, S. 2017. "A review on biogenic synthesis of ZnO nanoparticles using plant extracts and microbes: a prospect towards green chemistry." *Journal of Photochemistry and Photobiology B: Biology* 166: 272-284.

[73] Ogunyemi, S. O., Abdallah, Y., Zhang, M., Fouad, H., Hong, X., Ibrahim, E., ... and Li, B. 2019. "Green synthesis of zinc oxide nanoparticles using different plant extracts and their antibacterial activity against *Xanthomonas oryzae pv.* oryzae." *Artificial cells, nanomedicine, and biotechnology* 47(1): 341-352.

[74] Quek, J. A., Sin, J. C., Lam, S. M., Mohamed, A. R., and Zeng, H. H. 2020. "Bioinspired green synthesis of ZnO structures with enhanced visible light photocatalytic activity." *Journal of Materials Science: Materials in Electronics* 31: 1144-1158.

[75] Ibrahim, R. K., Hayyan, M., Al Saadi, M. A., Hayyan, A., and Ibrahim, S. 2016. "Environmental application of nanotechnology: air, soil, and water." *Environmental Science and Pollution Research* 23(14): 13754-13788.

[76] Baker, S., Volova, T., Prudnikova, S. V., Satish, S., and Prasad, N. 2017. "Nanoagroparticles emerging trends and future prospect in modern agriculture system." *Environmental toxicology and pharmacology* 53: 10-17.

[77] Rout, G. R., and Das, P. 2009. "Effect of metal toxicity on plant growth and metabolism: I. Zinc. In *Sustainable agriculture* (pp. 873-884)." Springer, Dordrecht.

[78] Raliya, R., Nair, R., Chavalmane, S., Wang, W. N., and Biswas, P. 2015. "Mechanistic evaluation of translocation and physiological impact of titanium dioxide and zinc oxide nanoparticles on the tomato (*Solanum lycopersicum L.*) plant." *Metallomics* 7(12), 1584-1594.

[79] Liu, R., and Lal, R. 2015. "Potentials of engineered nanoparticles as fertilizers for increasing agronomic productions." *Science of the total environment* 514: 131-139.

[80] Luque, P. A., Soto-Robles, C. A., Nava, O., Gomez-Gutierrez, C. M., Castro-Beltran, A., Garrafa-Galvez, H. E., Vilchis-Nestor, A. R., and Olivas, A. 2018. "Green synthesis of zinc oxide nanoparticles using *Citrus sinensis* extract." *Journal of Materials Science: Materials in electronics* 29: 9764-9770.

[81] Tettey, C. O., and Shin, H. M. 2019. "Evaluation of the antioxidant and cytotoxic activities of zinc oxide nanoparticles synthesized using *Scutellariabaicalensis* root." *Scientific African* 6: e00157.

[82] Ullah, S., Ahmad, A., Ri H., Khan, A. U., Khan, U. A., and Yuan, Q. 2019. "Green synthesis of catalytic Zinc Oxide nano-flowers and their bacterial infection therapy" *Applied Organometallic Chemistry* 34 (1): e5298.

[83] Kahsay, M. H., Tadesse, A., Rama, Devi D., Belachew, N., and Basavaiah, K. 2019. "Green synthesis of zinc oxide nanostructures and investigation of their photocatalytic and bactericidal applications." *RSC Advances* 9: 36967- 36981.

[84] Balaji, S., and Mandal, B. K. 2017. "Facile green synthesis of zinc oxide nanoparticles by *Eucalyptus globulus* and their photocatalytic and antioxidant activity" *Advanced Powder Technology* 28 (3): 785-797.

[85] Alamdari, S., Sasani Ghamsari, M., Lee, C., Han, W., Park, H. H., Tafreshi, M. J., Afarideh, H. and Ara, M. H. M., 2020. "Preparation and characterization of zinc oxide nanoparticles using leaf extract of *Sambucus ebulus*." *Applied Sciences*, 10(10): 3620.

[86] Varadavenkatesan, T., Lyubchik, E., Pai, S., Pugazhendhi, A., Vinayagam, R. and Selvaraj, R., 2019. "Photocatalytic degradation of Rhodamine B by zinc oxide nanoparticles synthesized using the leaf extract of *Cyanometra ramiflora*." *Journal of Photochemistry and Photobiology B: Biology*, 199: 111621.

[87] Kohane D. S. 2006. *Microparticles and nanoparticles for drug delivery. Biotechnology and Bioengineering* (Wiley).

[88] Murali, M., Kalegowda, N., Gowtham, H. G., Ansari, M. A., Alomary, M. N., Alghamdi, S., Shilpa, N., Singh, S. B., Thriveni, M. C., Aiyaz, M. and Angaswamy, N., 2021. "Plant-mediated zinc oxide nanoparticles: advances in the new millennium towards understanding their therapeutic role in biomedical applications." *Pharmaceutics*, 13(10): 1662.

[89] Vimala, K., Sundarraj, S., Paulpandi, M., Vengatesan, S., and Kannan, S. 2014. "Green synthesized doxorubicin loaded zinc oxide nanoparticles regulates the Bax

and Bcl-2 expression in breast and colon carcinoma." *Process biochemistry* 49(1): 160-172.

[90] Akbarian, M., Mahjoub, S., Elahi, S. M., Zabihi, E., and Tashakkorian, H. 2020. "Green synthesis, formulation and biological evaluation of a novel ZnO nanocarrier loaded with paclitaxel as drug delivery system on MCF-7 cell line." *Colloids and Surfaces B: Biointerfaces* 186: 110686.

Chapter 3

Biomolecules in Plant Extracts as Reducing Agents for AgNPs Synthesis

Yared Gutiérrez-Pinzón[1], MD
Gilberto Velázquez-Juárez[1], PhD
Edgar José López-Naranjo[2], PhD
and Diego Alberto Lomelí-Rosales[1,*], PhD

[1]Departamento de Química, Centro Universitario de Ciencias Exactas e Ingenierías, Universidad de Guadalajara, Guadalajara, Jalisco, México
[2]Departamento de Ingeniería de Proyectos, Universidad de Guadalajara, Zapopan, Jalisco, México

Abstract

Silver nanoparticles (AgNPs) are being widely studied due to their physical and chemical properties, which depend mainly on morphology, size, and composition, where the stabilizing agent is of great importance (Donahue et al. 2022).

When talking about obtaining AgNPs, several synthesis protocols are established, which include chemical (electrochemical, photochemical, and chemical reductions, among others) and physical methods (mainly laser ablation).(Iravani et al. 2014)However, derived from the interest in giving biomedical applications to AgNPs (antimicrobial and antioxidant activities or biosensing), a series of new methodologies have emerged called « green synthesis methods,» where biopolymers, biomass, plant extracts, and other varieties of natural products are used. Green synthesis methods can produce a new

---

* Corresponding Author's Email: diego.lomeli4077@academicos.udg.mx.

In: The Future of Green Synthesis
Editor: Jacques Kent
ISBN: 979-8-88697-816-2

generation of AgNPs with low toxicity and cost (Oceguera-Galindo, Oceguera-Contreras, and Pozas-Zepeda 2020).

The main objective of this review is to present the information derived from various investigations where plant extracts function as reducing agents for AgNPs, emphasizing the phytochemical composition of natural plant extracts. Biomolecules, naturally found in plant extracts, are responsible for reducing the metal ions in solution forming the AgNPs, and subsequently, some compounds can stabilize the AgNPs. Among the biomolecules studied with reducing capacity are carbohydrates, phenolic acids, flavonoids, proteins, and free amino acids. These molecules possess specific chemical characteristics that make them essential for understanding the mechanisms of AgNPs formation. Likewise, the pros and cons of the various synthetic methodologies and the biomedical applications presented by this new green generation of AgNPs are widely discussed.

Keywords: silver nanoparticles, biomolecules, green synthesis, plant extracts

Introduction

Nanotechnology is a field of study with great potential, which has been overgrown in recent years. Within the objectives of its areas are the design, synthesis, and manipulation of matter at the nanometric scale. Nanomaterials have physical, chemical, and biological properties that make them unique for new applications or to replace existing materials (Borse, Konwar, and Srivastava 2020).

Nanobiotechnology is an important research field that is rapidly growing because it is an interphase of nanotechnology and biotechnology, considering the application of nanomaterials in life science. Thus, one of the topics of research in nanobiotechnology is to explore new methodologies for synthesizing metallic nanoparticles (MNPs) with different chemical compositions, specific control of sizes and shapes, and controlled dispersity (Iravani et al. 2014; Alshora, Ibrahim, and Alanazi 2016; Kaur and Thombre 2021; K. K. Jain 2011).

Among the recently risen methodologies are the so-called green methodologies, which employ natural sources of reductants and stabilizers obtained from bacteria, plants, and fungi components, among others, to synthesize MNPs. This approach represents an alternative to developing eco-friendly, inexpensive, easy-to-use, and non-toxic MNPs. Within these green methods, also called biological methods, some relatively new terms, like phytosynthesis or phytoproduction, refer to using bioactive molecules extracted from plants (Rai, Yadav, and Gade 2008; Ahmad et al. 2022; Chandrasekharan, Chinnasamy, and Bhatnagar 2022; Husen and Siddiqi 2014).

Plants offer the benefits of being accessible, non-pathogenic, safe to handle, and rich in active agents. The main compounds of interest for synthesizing AgNPs present in plants include polysaccharides, terpenoids, alkaloids, flavones, amino acids, proteins, pigments, and vitamins, among others. Green synthesis using biological molecules derived from plant sources in the form of extracts is superior to chemical and other biological methods since they offer short-duration processes that can be performed even at room temperature showing highly controlled assembly that produces stable and shaped-controlled AgNPs (Ahmed et al. 2016; Lomelí-Rosales et al. 2022; Rafique et al. 2017).

However, despite all the advantages mentioned above of plant extract based routes, some major drawbacks include the slow rate of synthesis and the limited number of sizes and shapes of nanoparticles. Besides, the quality and the quantity of the biomolecules involved in the synthesis process can be affected by some environmental and nutritional factors such as light, temperature, and salinity among others, which affect the strength of the reducing agent. On the other hand, the size, morphology, and stability of the synthesized AgNPs can be optimized according to the control of some parameters such as the solvent, pH, and temperature established for the plant extraction, and concentration of the plant extract used (Ahmed et al. 2016, Rafique et al. 2016, Siddiqi, Husen, and Rao 2018, Qamer et al. 2021, Zahoor et al. 2021, Arif and Uddin 2020, Shanmuganathan et al. 2019, Srikar et al. 2016).

Conventional Methods versus Green Routes

Among all noble metallic nanoparticles, silver nanoparticles (AgNPs) have gained worldwide interest due to their unique set of properties such as chemical stability, good conductivity, catalytic, antibacterial, antiviral, and antifungal activities that can be used in the development of composite materials, cosmetics, food products, and electronic components. AgNPs can be obtained through different routes, which are classified into three mains groups: physical, chemical, and biological, each one of them encompassing a different group of techniques (Table 1) (Ahmed et al. 2016, Alvarado-

Arámburo et al. 2022, Hilal, Mohammed, and Shakir 2019, Issa et al. 2017, Rafique et al. 2016).

Physical routes follow the "top-down" approach, which works with the material in bulk and splits it into fine particles by size reduction by specialized ablation methods. Although this type of process produces AgNPs with a narrow size distribution, they are also characterized by employing expensive equipment and a large amount of energy consumption, consequently raising production costs.

Besides, the use of polluting solvents, low productivity, and products showing a lack of uniform distribution and surface structural defects are common drawbacks, making this group of methods less attractive. Such surface defects significantly negatively impact the physical features and surface chemistry of the metallic nanoparticles. On the other hand, chemical and biological methods follow the "bottom-up" approach, meaning that the AgNPs are obtained by the self-assembly phenomenon of atoms to nuclei.

Chemical methods involve the reduction of silver salts and nucleation, followed by the growth of the MNPs. These methods are the most widely used due to the significant amount of MNPs that can be obtained in short periods. However, in chemical methods, organic and inorganic compounds are employed to reduce Ag^+ ions that show hazardous byproducts that negatively affect the environment and human health.

Thus, biological methods, which do not employ toxic chemicals, constitute an alternative option to obtain AgNPs through plant biomass, plant extracts, or microorganisms. These methods are inexpensive, dependable, environmentally friendly, fast, and simple. Different options have been evaluated regarding biological methods, as indicated in Table 1. For example, bacterial species have been used in the synthesis of AgNPs due to their inherent ability to reduce heavy metals, as the bacteria's cell wall is responsible for the reduction due to the presence of various functional groups. Bacteria-mediated synthesis constitutes a flexible, reasonable, and suitable option for large-scale synthesis but at a slow rate and with a limited number of sized and shaped NPs.

For example, actinomycetes (a group of bacteria) have been successfully used because the enzymes secreted from their cell wall and membrane help reduce Ag^+ ions, and their proteins help in capping and stabilizing the AgNPs. On the other hand, fungi-based methods have drawn interest mainly due to economic reasons and potential applicability in different fields. Fungi produce enzymes and proteins that reduce and stabilize metal nanoparticles; compared to bacteria, the nanoparticles produced by this fungi-based method are higher, so this process shows the potential for large-scale production of AgNPs. In the meantime, AgNP synthesis by plant based-routes takes advantage of the diverse metabolites presented in different parts of plants (e.g., leaves, roots, fruits, peel, bark, flowers, and sprouts), which have been used to obtain compounds employed as capping, stabilizing, and reducing agents.

Table 1. Chemical, physical and biological methods for the synthesis of AgNPs (Rafique et al. 2016)

Chemical	Physical	Biological*
Chemical reduction	Pulsed laser ablation	Bacteria
Sonochemical	Evaporation-condensation	Fungi
Photochemical	Arc discharge	Plant extracts
Microemulsion	Spray pyrolysis	Yeast
Electrochemical	Ball milling	Enzymes
Pyrolysis	Vapor and gas phase	Actinomycetes
Microwave	Pulse wire discharge	DNA
Solvothermal	Lithography	Viruses

*Based on the use of one of the different options enlisted.

Active Biomolecules Found in Plant Extracts for the Synthesis of AgNPs

Plants present a wide diversity of compounds that can act as reducing agents in the synthesis of nanoparticles; for example, primary metabolites such as simple and complex carbohydrates, free amino acids, and proteins. All of them have been reported as reducing agents for the formation of nanoparticles. In addition, it is not only primary metabolites that can fulfill this function. Plants possess several compounds that serve as defense molecules against predators and pathogens, known as secondary metabolites. Plants' most reported secondary metabolites include polyphenols, tannins, flavonoids, and stilbenes. Since each plant produces a different set of secondary metabolites depending on the stress to which it is subjected, the myriad of compounds obtained from secondary metabolism is immense. That is why multiple reports of nanoparticle synthesis are based on phytochemical extracts containing secondary metabolites. One family of chemical compounds, from secondary metabolism in particular, has been of significant importance in recent publications, namely polyphenols. Polyphenols have been reported to be mainly responsible for the synthesis of nanoparticles in extracts used in

AgNPs photosynthesis. However, they are also possibly responsible for some of certain nanoparticles' antioxidant and antimicrobial activities (AAM et al. 2018; Koduru et al. 2018; Olszowy 2019; Vuolo, Lima, and Marostica Junior 2019; Khan, Shahid, and Lee 2020).

The reduction of Ag^+ ions depend on the active functional groups presented by the extract's biomolecules; - OH, -CHO, $-NH_2$, and phenols are the most common functional groups involved in the Phyto production of AgNPs (Koduru et al. 2018). Within the set of components found in the plant extract, the most commonly reported for the synthesis of AgNPs are polyphenols, flavonoids, tannins, and amino acids, among others. An interesting behavior of the biomolecules present in the plant extract is that they can achieve AgNPs synthesis in a single step (Lomelí-Rosales et al. 2022; Yousaf et al. 2020). In addition, phytochemical compounds can be anchored to the surface of AgNPs, generating a protective and stabilizing layer on the surface of AgNPs, and preventing their agglomeration (Koduru et al. 2018). In other words, those biomolecules present a synergistic effect on the surface of the AgNPs, acting as reductants and stabilizer agents (capping agents). Theses biomolecules on the AgNPs surface are the ones that can increase the antioxidant and antimicrobial activity of the AgNPs (Yousaf et al. 2020).

The nanoparticles synthesized using plant extracts exhibit physicochemical characteristics that are closely dependent on the phytochemical profile present in the extract (Koduru et al. 2018). The phytochemical profile of an extract depends on various factors, both biological (abiotic and biotic factors present during plant development, as well as the plant tissue used to obtain the extract) as well as methodological factors (treatment of the tissue to carry out the extraction, type of solvent used, extraction method, extraction time, and temperature). For example, the polarity of the solvent used will determine the type of molecules that can be obtained in the extract. In this sense, for example, flavonoids and phenolic compounds can be found in mayor concentration when a non-polar or low-polar solvents, and in less concentration with polar solvents are used for the extraction (Akinmoladun et al. 2022; Nawaz, Aslam, and Muntaha 2019; Nawaz et al. 2020). Various extraction methodologies have been reported to obtain the families of organic compounds (biomolecules) for the green synthesis of AgNPs. Some of the reported strategies are described below.

In the research carried out by Swilam and co-workers, the polyphenols, and flavonoids profile in pomegranate leaves in three different fractions extract (ethyl acetate, crude extract and n-butanol) was reported. They found, ethyl acetate fraction was the richest fraction in flavonoids and polyphenols, followed by the crude extract, and the lowest concentration was found in n-butanol fraction. When analyzing the ethyl acetate fraction by LC-MS-MS (liquid chromatography-mass spectrometry), 23 compounds were found: 11 compounds were identified as phenolic acids and their derivatives, 8 turned out to be tannins (7 hydrolyzable tannins and 1 condensed tannin), 3 anthocyanins and 1 flavonoid derivate. After the identification of the biomolecules in the extracts, the synthesis of AgNPs proceeded, finding that the ethyl acetate fraction led to obtaining the AgNPs with better characteristics, compared to the other two fractions (Swilam and Nematallah 2020).

The phytochemical analysis in the investigation reported by Jalilian and co-workers for *Allium ampeloprasum* showed the presence of a group of several bioactive molecules as organic acids, phenolic compounds, aliphatic disulfides, aliphatic amines, and amino acids when water was used as solvent extraction and the mixture was boiled for 15 minutes (Jalilian et al. 2020). A similar procedure where *Capsicum chinense* aqueous extract was prepared at 80°C for 20 minutes and used for the preparation of AgNPs, the authors found in the FTIR (Fourier Transformation Infrared) that reductant sugar, free amino acid, and phenols were the responsible bioactive molecules for the stability of the obtained AgNPs (Lomelí-Rosales et al. 2022). On the other side, Behboodi and co-workers also found terpenoids and phenol groups as responsible molecules for the bio-reduction and fabrication of AgNPs when Cichorium intybus aqueous extract was used (Behboodi et al. 2019).

Mat Yasulf and co-workers studied the *Clinacanthus nutans* extracts (leaf and steam). For the preparation of the extracts, the authors took into consideration the thermolabile compounds of the plant, for which they use the maceration method, and distilled water was the chosen extraction solvent. The phytochemical analysis showed the presence of tannins, saponins, alkaloids, amino acids, flavonoids, reductant sugars, and glycosides. The analysis of the extract and the AgNPs obtained, by FTIR, showed the presence of -OH and -NH_2 groups, which must correspond to phenolic and amide compounds respectively, and they could be involved in the Ag+ ions reduction for the AgNPs synthesis (Mat Yusuf et al. 2020).

Alshehri and Malik reported a complete study of *Matricaria chamomilla* extracts for the photo-induced green synthesis of AgNPs. The authors found, by FTIR analysis, that the -OH bonds of the exposed flavonoids to bright sunlight reduce the Ag^+ ions due to the oxidation of this group that converts the phenol to the stable keto form in apinenin-7-O-glucoside, apigenin, quercetin, and luteolin (Alshehri and Malik 2020). The phytochemical analysis of this plant demonstrated that it possesses sesquiterpenes, polyacetylenes, and coumarins, groups of molecules that can contribute to both, synthesis and capped AgNPs.

Meanwhile, Yousaf and co-workers used aqueous and alcoholic extracts of *Achillea millefolium* for the synthesis of AgNPs. According to the FTIR, they proposed that phenols (such as tannic acid), flavonoids, and carboxylic acids (proteins) were the responsible biomolecules involved in the formation of the nanoparticles (Yousaf et al. 2020). In another study conducted by Wang and co-workers using *Psidium guajava* extract as a source of reducing and capped biomolecules, the presence of flavonoids and polyphenols was also found by FTIR analysis (Wang et al. 2018).

Chemical Characterization Techniques Use in Green Synthesis of AgNPs

As can be read in the previous paragraphs, in addition to the extraction method for obtaining biomolecules that conduce the synthesis of AgNPs, a topic of current interest in this field, is the chemical analysis technique used with the aim of elucidating the groups of molecules or the specific compound present in the plant extract. Several methods of chemical analysis are being applied to elucidate the composition of the extracts and also for the molecules that actually intervene in both, the synthesis and stability of the AgNPs.

One of the most widely used methods for monitoring is Fourier-transform infrared spectroscopy (FTIR). In that method, it is possible to know the composition of functional groups on the surface of nanoparticles, as well as the phytochemical compounds involved in the synthesis and stabilization of AgNPs (Koduru et al. 2018). FTIR covers the vibrational frequencies (4000 to 400 cm-1) of many of the chemical groups present in almost all organic and inorganic compounds, which facilitates the analysis of both AgNPs, but also of the extracts used for their synthesis (Tatulian 2019). On the other hand, UV/VIS spectrometry is routinely used for the characterization of nanoparticles immediately after their synthesis. Since it is a simple and inexpensive method, it also allows us to evaluate at a first glance the composition of the extract. It is recommended to perform an initial spectrum of the extract, the supernatant, and the purified nanoparticle solution, to observe that the absorbance spectrum does not overlap one with the other, avoiding possible reading errors or inconsistencies in future analyses (De

Menezes et al. 2021a; Koduru et al. 2018). Although there are more chemical techniques applied in these areas, these are two of the used in all cases.

Yallappa and co-workers prepared a seed extract from *Acacia farnesaiana*. The extraction was obtained in water as dissolvent and microwave radiation for 180 s. The prepared AgNPs were characterized by spectroscopy techniques: FTIR showed the presence of amide groups, C-N from amine, which corresponds to flavanones and terpenoids present on the AgNPs surface; and ^{13}C NMR (^{13}C Nuclear Magnetic Resonance) spectrum, which presented three groups of signals: the corresponding for ketones and aldehydes (flavanones, terpenoid aldehydes), the amide group signal, unsaturated compound, and sugars. The samples were also analyzed by the energy dispersion X-ray (EDX) technique for obtaining elemental composition and by the AgNPs, founding the evidence of C and O in the sample, which corresponds to organic substances (Yallappa et al. 2013).

In some reports, the isolation of some components of the plant extract was possible due to the high concentration of the compound and the extraction method used. For example, Swilam and Nematallah reported the isolation of the major phenolic compound from *Punica granatum*, its components were characterized by ^{1}H-RMN spectroscopy, and they concluded that the signals belong to ellagic acid (Swilam and Nematallah 2020).

Recently, others several techniques have been used that together allow us to determine the phytochemical profile involved in the synthesis of AgNPs and their stabilization. A brief example is that of Pradeep and collaborators (2022), who, with the support of liquid chromatography coupled to a photodiode array and a mass spectrophotometer, were able to determine that in the case of AgNPs synthesized with *Hypericum perforatum* extracts the synthesis of AgNPs occurs mainly by the action of phenolic acids and flavonoids, and stabilization occurs to a greater extent by xanthones and fluoroglycerols (Pradeep et al. 2022b).

On the other hand, Pradeep and co-workers reported a phytochemical analysis of *Hypericum perforatum* extract and for the postreaction residues carried out by ultra-performance liquid chromatography coupled with a photodiode array and high-resolution mass spectrometry (UPLC-PDA-HRMS). The results showed the presence of 41 compounds in aqueous extract and after the synthesis of AgNPs authors found that phenolic acids and flavonoids were involved in the reduction of $Ag+$ ions (chlorogenic acid, quercetin, protohypericin, hypericin, pseudohypericin), the xanthones and phloroglucinols were the responsible for the stabilization of the AgNPs (paxanthonin, paxathona, calycinoxanthone D, adhyperforin, furohyperforin), and the naphthodianthrones were involved in both steps (Pradeep et al. 2022a).

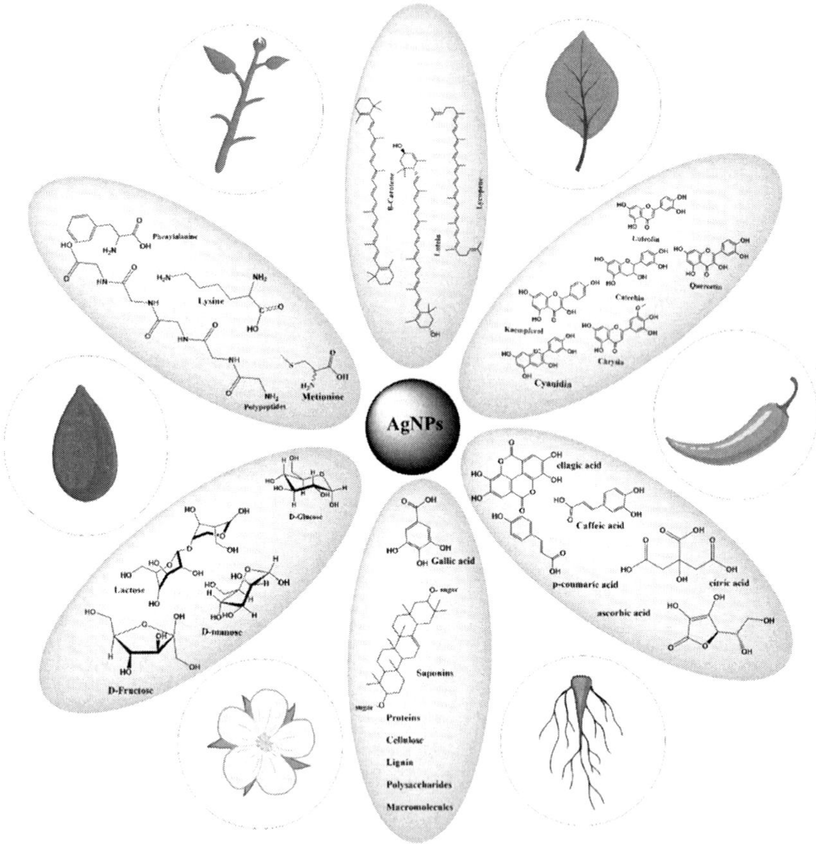

Figure 1. Representation of part of plants and some of the common biomolecules (groups of compounds) found in plant extracts reported for AgNPs synthesis and stabilization.

A "green" synthesis using basic amino acids (L-lysine or L-arginine) as reducing agents and soluble starch as a stabilizer was reported by Hu and co-workers. The results of this investigation were the formation of semi-spherical AgNPs. In this study was demonstrated that amino acids are possible molecules that can conduct the reduction of Ag^+ ions for the formation of AgNPs (Hu et al. 2008). Similar research was carried out using eosin methylene blue agar (EMB) as a source of reducing and stabilizer agents, and a comparison was made with the pure components of the agar (lactose, sucrose, peptone, salts, eosin, methylene blue and agar) for the formation of AgNPs. AgNPs were not only successfully obtained using EMB, but also with the used of lactose and sucrose, as pure components. The authors demonstrate that lactose and sucrose were necessary for the reduction of the Ag+ ions and eosin and methylene blue did not intervene in the synthesis or even in the stabilization of the AgNPs (Lomelí-Rosales et al. 2019). Even though these reports did not use a plant extract, is useful to compare one bio active group of molecules (amino acids and sugars, among others) that are commonly found in plants as possible reductant / stabilizers agents for metals. In this sense, Jain and Mehata employed the leaf extract of *Ocicum sanctum* and its pure derivate quercetin (a flavonoid present in the plant) for the synthesis of AgNPs. They found that the quercetin allowed the formation and stabilization of AgNPs, so they concluded that flavonoids contained in a plant are mainly responsible to the formation of AgNPs (S. Jain and Mehata 2017).

Chemical Methods to Evaluate the Presence and Antioxidant Capacity of Phytochemical Compounds

To have a better understanding of the compounds involved in the synthesis and stabilization of AgNPs, it is required to evaluate the phytochemical profile of the extracts used, the AgNPs synthesized, as well as the supernatant obtained during the purification of the AgNPs. Some of the tests mostly used to evaluate the phytochemical profile of the nanoparticles synthesized using plant extracts are briefly described below.

In general, antioxidant activity can be described as the ability of some compounds to reduce reactive species, such as free radicals (Olszowy and Dawidowicz 2018; Abuelizz et al. 2020). Two types of mechanisms by which the chemical reaction between the free radical and the antioxidant agent occurs have been described. The first one is the HAT (Hydrogen Atom Transfer) mechanism, HAT-type methods measure the ability of an antioxidant to stabilize the free radical by donating hydrogen, see equation (1). These types of reactions are usually very fast (seconds to minutes), and the reaction is usually not pH-dependent (AAM et al. 2018).

$$X^{\bullet} + AH \rightarrow XH + A^{\bullet} \tag{1}$$

The second mechanism is the SET (Single Electron Transfer) type. In this type of method, the ability of an antioxidant to reduce a free radical by donating an electron is detected. This type of reaction is pH-dependent and, unlike the HAT-type methods, requires a longer incubation time for the reaction to take place.

$$X^{\bullet} + AH \rightarrow H^{-} + AH^{+\bullet} \quad (2)$$

Colorimetric methods for the evaluation of antioxidant activity are the most widely used, due to their simplicity, short experimental time, and do not require any complex equipment for analysis, only a UV/VIS spectrophotometer (Olszowy and Dawidowicz 2018; Gupta, Caraballo, and Agarwal 2018).

In this type of method based on colorimetry, it is recommended to monitor that the extracts, supernatants, and nanoparticles to be evaluated do not present absorbance peaks in the wavelength range of the method used to evaluate the activity, since reading errors may occur. It is recommended to perform the reading at another wavelength of the radical color, whenever possible (Olszowy and Dawidowicz 2018; de Menezes et al. 2021).

DPPH (2-2-Diphenyl-1-Picrylhydrazyl)

$DPPH^{\bullet}$ inhibition assessment is the mostly used quantification method in the evaluation of silver nanoparticle antioxidant activity. It is based on the quantification of the absorbance change, stability, and coloration of the $DPPH^{\bullet}$ radical (Olszowy and Dawidowicz 2018). It is a stable radical soluble in organic solvents such as methanol, ethanol, chloroform, ether, acetic acid, or in a mixture of these with water, provided that the percentage of water is not more than 60%. $DPPH^{\bullet}$ absorbs intensely in the visible region of 513-528 nm, and exhibits a purple coloration, while its reduction in the presence of antioxidants produces a yellowish coloration (Nenadis and Tsimidou 2018; Olszowy and Dawidowicz 2018 Nenadis and Tsimidou 2018).

Traditionally, the mechanism of radical inhibition has been described as an exchange of a hydrogen atom from the antioxidant compound to the $DPPH^{\bullet}$ radical, to produce DPPH2. Although it has been shown that the reaction of the antioxidant with $DPPH^{\bullet}$, can be of the HAT or SET type. Commonly, a solution of known concentration of TROLOX (6-hydroxy-2,5,7,8-tetramethyl chroman-2-carboxylic acid), which is an analogue of vitamin E, is used as a standard for quantification. However, other antioxidants have also been used, such as 2-carboxylic acid, ascorbic acid, catechin, caffeic acid, and gallic acid (Abramovič et al. 2018).

Some of the advantages of this method are its simplicity, low cost, the operation skills are not very strict, and it does not require very specialized technology (de Menezes et al. 2021).

One of the most used ways to express the results of this method is by the percentage of DPPH scavenging activity:

$$\%\text{DPPH scavenging activity} = \frac{OD_{control} - OD_{sample}}{control} \text{x } 100 \quad (3)$$

ABTS 2,2'-Azinobis (3-Ethylbenzothiazoline-6-Sulfonic Acid)

The generation of $ABTS^{\bullet+}$ monocation is produced by the reaction between the ABTS solution and some initiator, such as metmyoglobin /H_2O_2, potassium or ammonium persulfate, manganese dioxide, etc., which produces a blue-green solution, resulting in a blue-green solution, this coloration is lost when $ABTS^{\bullet+}$ reacts with some antioxidant compound, so the result is a pale yellow-crystalline solution (Ilyasov, Beloborodov, and Selivanova 2018; Kut et al. 2022). The reaction mechanism is both HAT and SET, depending on the type of antioxidant with which it reacts, exhibits an absorbance between 414-417 nm and 730-734 nm.

Unlike other radicals, ABTS is soluble in both water and other organic solvents (methanol, ethanol, acetic acid), but also in buffers used in biological analysis such as phosphate buffer (PBS). Which facilitates its use for the evaluation of the antioxidant activity of AgNPs intended to be used in a biological system (Zheng et al. 2016).

A Trolox solution of known concentration is usually used as a standard, so this assay is also referred to as Trolox equivalent antioxidant capacity (TEAC). Although other antioxidants can also be used as a standard. This test is stable over a range of 2.0-10.5 pH, although it is more stable at acidic pH (Ilyasov et al. 2020; Kut et al. 2022). Moreover, it is a popular, simple, and highly reproducible method (Ilyasov, Beloborodov, and Selivanova 2018).

The results of this test can be expressed as percentages of $ABTS^{\bullet+}$ radical inhibition:

$$\% \text{ ABTS}^{\bullet+} \text{ scavenging activity} = \frac{OD_{control} - OD_{sample}}{control} \text{x } 100 \quad (4)$$

FRAP (Ferric Reducing Antioxidant Power)

It is a method based on electron transfer (SET). Iron in the oxidized form Fe^{3+} is reduced to Fe^{2+} in an aqueous solution at acidic pH (acetate buffer). The solution of the Fe^{3+} complex with 2 ligands of TPTZ (tris(2-pyridyl)-S-triazine) is yellow, when the reduction reaction of Fe^{3+} to Fe^{2+} by the antioxidant compound occurs, the coloration changes to an intense blue, presenting an absorbance at 593 nm after an incubation period of 30 to 90 minutes. The working solution for this method is a mixture of TPTZ, HCl, and $FeCl_3$. A Trolox solution with a known concentration is usually used as a standard (Gupta, Caraballo, and Agarwal 2018; Sinan et al. 2021; Soural et al. 2022; Xiao et al. 2020). As can be deduced the pH value is very important for the reaction to take place, so the pH of the extracts to be evaluated should be considered since in solutions of basic pH an increase in antioxidant activity can be observed because protons can dissociate from the phytochemical compounds (mainly phenolics), and therefore, the readings are not reliable (Vuolo, Lima, and Maróstica Junior 2019).

Less frequently, other methods have been used to evaluate antioxidant activity in AgNPs using chemical methods such as inhibition of radicals: H_2O_2, HO y O_2 (Valgimigli, Baschieri, and Amorati 2018; Xiao et al. 2020).

Quantification of Total Polyphenols by the Folin-Ciocalteu Method

The method most commonly used to quantify the number of polyphenols is the Folin-Ciocalteu assay, which is based on the SET-type reaction of phenolic compounds with molybdenum in an alkaline solution, when the reaction occurs the working solution turns from a yellowish-crystalline to an intense blue color. The measurement is carried out in the absorbance spectrum between 750 and 765 nm. Gallic acid at a known concentration is usually used as a standard. However, other antioxidants such as caffeic acid, tannic acid, and catechin can also be used. The working solution can be found commercially (Phuyal et al. 2020; Wong and Tan 2020).

In addition to the quantification of total phenolic compounds, it is also suggested to carry out the determination of reducing sugars, total proteins, and free amino acids present in the extracts, AgNPs, and their supernatant. Table 2 summarize some of the recent investigations where the antioxidant activity has been carried out for plant extracts, or an extract of one part of the plant, used in the synthesis of nanoparticles.

Table 2. Compilation of some reported studies of plant extracts using for the formation of AgNPs: solvent extraction, methodology for determination of antioxidant capacity, size, and shape obtained

Species	Tissue	Extraction solvent	Antioxidant Method	AgNP size	AgNP shape	Reference
Achillea millefolium L.	All plant	Aqueous; methanol; ethanol	DPPH	20.77 nm	Spherical	(Yousaf et al. 2020)
				18.53 nm	Rectangular	
				14.27 nm	cubical	
Allium ampeloprasum	All plant	Aqueous	DPPH	8-50 nm	Spherical	(Jalilian et al. 2020)
Brassica oleracea	Leaves	Aqueous	DPPH; Superoxide radical	20 nm	Spherical	(Ansar et al., 2020)
Capsicum chinense	Steam, root, and leaves	Aqueous	ABTS, DPPH, and FRAP.	20.67 nm	Spherical	(Lomelí-Rosales et al. 2022)
Carissa carondas L.	Leaves	Aqueous	DPPH	30-35 nm	Cubical	(Singh et al. 2021)
Cassia angustifolia	Flower	Aqueous	DPPH; Hydrogen peroxide; FRAP	10-80 nm	Spherical	(Bharathi and Bhuvaneshwari 2019)
Cestrum nocturum	Leaves	Aqueous	DPPH; Hydrogen peroxide; Hydroxyl radical; Superoxide radical	20 nm	spherical	(Mousavi-Khattat, Keyhanfar, and Razmjou 2018)
Datura stramonium	Leaves	Aqueous	DPPH	13-60 nm	Spherical	(Mousavi-Khattat, Keyhanfar, and Razmjou 2018)
Diospyros Montana	Steam bark	Methanol	DPPH; Hydrogen peroxide	28 nm	Spherical	(Bharathi et al. 2018)
Euphorbia helioscopia	Leaves	Aqueous	Superoxide dismutase assay; Ascorbate Peroxidase assay	30-100 nm	Spherical	(Batool et al. 2021)
Moringa oleifera	Leaves	Aqueous	DPPH	23.5- 53.4 nm	Spherical	(Mohammed et al. 2022)
Mussaenda frondosa	Leaves	Aqueous	DPPH	30-60 nm	Rod, triangle and quasi-spherical	(Sreelekha et al. 2021)
Plantago lancerolata	All plant	Methanol	DPPH	20-30 nm	Spherical	(Shah et al. 2021)
Psidium guajava L.	Leaves	Aqueous	DPPH; $ABTS^{+}$	20-35 nm	Spherical	(Wang et al. 2018)
Tropaeolum majus L.	Leaves	Aqueous	DPPH; $ABTS^{+}$;	35-55 nm	Spherical	(Valsalam et al. 2019)
		Ethanol	Total antioxidant activity			

Applications of AgNPs

Due to their wide range of properties, AgNPs have been broadly used in different applications. AgNPs could help to overcome the bacterial resistance against antibiotics in the era of antibiotic resistance by replacing current antibiotics as alternative antibacterial agents or could also be used to improve the efficiency of certain antibiotics (e.g., the potential of amoxicillin, penicillin and others against *E. coli* and *S. aureus* has been observed to increase in the presence of green synthesized AgNPs).

Due to their wide variety of antimicrobial and antifungal applications, AgNPs have been used as antibacterial coatings in different products such as implants, dressing, dental composites, biosensors, and agriculture engineering. The antimicrobial properties of AgNPs depend on their size, shape, and environmental conditions (e.g., pH, ionic strength) as well as in the type of capping agent used in the synthesis process. NPs with same surface areas but dissimilar shapes exhibit differences in antibacterial activities.

Other factors such as stability (i.e., agglomeration and sedimentation) and aging of the sample are also important as an increase in the toxicity level has been reported for aged AgNPs stored in water for 6 months, which is related to the release of Ag+ ions. Although at present, the bacterial inhibition mechanism of AgNPs is still unclear, their antibacterial impact can be explained by three different hypotheses: contact action, the formation of reactive oxygen species (ROS), and the release of Ag+. It is known that AgNPs penetrate the bacterial cells inducing cell death. The entry of AgNPs into the bacterial cells causes DNA damages and modify their functions. The binding of Ag+ ions cause the breakdown of the bacterial cell walls disturbing the mechanism of protein synthesis. Moreover, AgNPs promote the production of ROS causing the inhibition of pathogenic bacteria. Plant extract-mediated synthesized AgNPs have also shown selective cell specific toxicity of cancer cells making them a potential method for cancer treatment.

Water purification and treatment is another area of potential application for AgNPs; so far, due to their high stability, cost effectiveness and controllable release, they have been successfully employed for the removal of inorganic anions, heavy metals, organic pollutants, and bacteria from water. In addition, AgNPs show potential applications in catalysis, plasmonic, optoelectronics, climate change, contamination control, and information storage. Due to their wide range of applications, a production of around 320 tons of AgNPs per year is estimated. Products prepared using AgNPs have been approved by a range of accredited bodies including the US FDA, US

EPA, Korea's Testing, SIAA Japan, Research Institute for Chemical Industry and FITI Testing and Research Institute (Ahmed et al. 2016, Rafique et al. 2016, Siddiqi, Husen, and Rao 2018, Qamer et al. 2021, Zahoor et al. 2021, Arif and Uddin 2020, Srikar et al. 2016).

Beyond all the advantages that AgNPs show, some potentially hazardous effects of AgNPs have caught worldwide attention. For example, they could constitute a potential threat of toxicity to aquatic organisms, and their extensive use as antimicrobial agents and disinfectants could enhance bacterial resistance (Zahoor et al. 2021).

Conclusion

All the above synthesis examples presented direct the "conventional" chemical reduction synthesis, to one whit a similar result minus the damaging repercussions that this entails, thus directing the production of AgNPs towards a "green" synthesis path, gentler with the environment.

The factor of achieving that within a plant extract, a single pure component can be isolated and used for the specifically directed synthesis of AgNPs is one of the challenges that arise in this area. Likewise, knowing the chemical structure of the component(s) that are present on the AgNPs surface for its stabilization could open more specific applications.

This field of biotechnology, for the synthesis of MNPs, is a multidisciplinary area, which too be fully understood requires a bunch of knowledge, for example, the chemistry of reducing agents (REDOX reactions), the chemistry of natural products (polarity of solvents and techniques for extractions), and the characterization techniques such as spectroscopy (FTIR, UV/vis, NMR, XPS), chromatography (HPLC, LC-MS-MS, GC-MS), and microscopy (TEM, SEM, EDX), to mention a few areas. Some other fields that are taken into consideration are called "omics sciences," especially metabolomics, which can be used to determine the phytochemical composition (primary and secondary metabolites) of plant extracts for both, polar and non-polar components.

In addition to the issues mentioned above, the phytosynthesis of nanoparticles requires more attention to the understanding and characterization of the biomolecules involved before and during the synthesis process. Understanding which are the groups of molecules and the proportions in which they are found while the synthesis is performed will allow to make the experiments more reproducible, verifiable, and efficient. Few publications indeed provide sufficient analysis of the biomolecules during the synthesis process, which can lead to the complication of adapting the conditions of the experiment each time it is attempted to be reproduced in another laboratory. On the other hand, the extraction efficiency and even the natural variability of the extracts due to the place where the plant material was obtained, undoubtedly generate a challenge to establish the experimental conditions for any laboratory other than the one where the nanoparticle was originally obtained. New metabolic profiling techniques, together with the established classical ones, will be fundamental to develop controlled and efficient processes for phytosynthesis.

References

Abramovič, Helena, Blaž Grobin, Nataša Poklar Ulrih, and Blaž Cigić. 2018. "Relevance and Standardization of in Vitro Antioxidant Assays: ABTS, DPPH, and Folin–Ciocalteu." *Journal of Chemistry* 2018. https://doi.org/10.1155/2018/4608405.

Abuelizz, Hatem A., El Hassane Anouar, Mohamed Marzouk, Hanan A. A. Taie, Adi Ahudhaif, and Rashad Al-Salahi. 2020. "DFT Study and Radical Scavenging Activity of 2-Phenoxypyridotriazolo Pyrimidines by DPPH, ABTS, FRAP and Reducing Power Capacity." *Chemical Papers* 74 (9): 2893–99. https://doi.org/10.1007/s11696-020-01126-0.

Ahmad, Bilal, Li Chang, Usama Qamar Satti, Sami ur Rehman, Huma Arshad, Ghazala Mustafa, Uzma Shaukat, Fenghua Wang, and Chunyi Tong. 2022. "Phyto-Synthesis, Characterization, and In Vitro Antibacterial Activity of Silver Nanoparticles Using Various Plant Extracts." *Bioengineering* 9 (12). https://doi.org/10.3390/bioengineering9120779.

Ahmed, Shakeel, Mudasir Ahmad, Babu Lal Swami, and Saiqa Ikram. 2016. "A Review on Plants Extract Mediated Synthesis of Silver Nanoparticles for Antimicrobial Applications: A Green Expertise." *Journal of Advanced Research* 7 (1): 17–28. https://doi.org/10.1016/j.jare.2015.02.007.

Akinmoladun, Afolabi C., Olubusola Elizabeth Falaiye, Olubukola Benedicta Ojo, Abimbola Adeoti, Zainab Abiola Amoo, and Mary Tolulope Olaleye. 2022. "Effect of Extraction Technique, Solvent Polarity, and Plant Matrix on the Antioxidant Properties of *Chrysophyllum albidum* G. Don (African Star Apple)." *Bulletin of the National Research Centre* 46 (1). https://doi.org/10.1186/s42269-022-00718-y.

Alshehri, Abdulmohsen Ali, and Maqsood Ahmad Malik. 2020. "Phytomediated Photo-Induced Green Synthesis of Silver Nanoparticles Using *Matricaria chamomilla* L. and Its Catalytic Activity against Rhodamine B." *Biomolecules* 10 (12): 1–24. https://doi.org/10.3390/biom10121604.

Alshora, Doaa Hasan, Mohamed Abbas Ibrahim, and Fars Kaed Alanazi. 2016. *Nanotechnology from Particle Size Reduction to Enhancing Aqueous Solubility.*

Surface Chemistry of Nanobiomaterials: Applications of Nanobiomaterials. Elsevier Inc. https://doi.org/10.1016/B978-0-323-42861-3.00006-6.

Ansar, S., Tabassum, H., Aladwan, N. S. M., Naiman Ali, M., Almaarik, B., AlMahrouqi, S., Abudawood, M., Banu, N., & Alsubki, R. (2020). Eco friendly silver nanoparticles synthesis by *Brassica oleracea* and its antibacterial, anticancer and antioxidant properties. *Scientific Reports*, *10*(1), 1–12. https://doi.org/10.1038/s41598-020-74371-8.

Batool, S. U., Javed, B., Sohail, Zehra, S. S., Mashwani, Z.-R., Raja, N. I., Khan, T., ALHaithloul, H. A. S., Alghanem, S. M., Al-Mushhin, A. A. M., Hashem, M., & Alamri, S. 2021. "Exogenous Applications of Bio-Fabricated Silver Nanoparticles to Improve Biochemical, Antioxidant, Fatty Acid and Secondary Metabolite Contents of Sunflower." *Nanomaterials* 11 (7): 1750. https://doi.org/10.3390/nano11071750.

Bharathi, Devaraj, and V. Bhuvaneshwari. 2019. "Evaluation of the Cytotoxic and Antioxidant Activity of Phyto-Synthesized Silver Nanoparticles Using *Cassia angustifolia* Flowers." *Bio Nano Science* 9 (1): 155–63. https://doi.org/10.1007/s12668-018-0577-5.

Bharathi, Devaraj, M. Diviya Josebin, Seerangaraj Vasantharaj, and V. Bhuvaneshwari. 2018. "Biosynthesis of Silver Nanoparticles Using Stem Bark Extracts of *Diospyros montana* and Their Antioxidant and Antibacterial Activities." *Journal of Nanostructure in Chemistry* 8 (1): 83–92. https://doi.org/10.1007/s40097-018-0256-7.

Borse, Vivek B., Aditya N. Konwar, and Rohit Srivastava. 2020. *Nanobiotechnology Approaches for Miniaturized Diagnostics. Handbook on Miniaturization in Analytical Chemistry: Application of Nanotechnology*. Elsevier Inc. https://doi.org/10.1016/B978-0-12-819763-9.00013-1.

Chandrasekharan, Smitha, Gandhimathi Chinnasamy, and Somika Bhatnagar. 2022. "Sustainable Phyto-Fabrication of Silver Nanoparticles Using *Gmelina arborea* Exhibit Antimicrobial and Biofilm Inhibition Activity." *Scientific Reports* 12 (1): 1–16. https://doi.org/10.1038/s41598-021-04025-w.

Donahue, Nathan D., Sandy Kanapilly, Chady Stephan, M. Caleb Marlin, Emmy R. Francek, Majood Haddad, Joel Guthridge, and Stefan Wilhelm. 2022. "Quantifying Chemical Composition and Reaction Kinetics of Individual Colloidally Dispersed Nanoparticles." *Nano Letters* 22 (1): 294–301. https://doi.org/10.1021/acs.nanolett.1c03752.

Elnour, A. A. M., Mohamed E. S. Mirghani, Musa K. H., Kabbashi N. A., and Alam M. Z. 2018. "Challenges of Extraction Techniques of Natural Antioxidants and Their Potential Application Opportunities as Anti-Cancer Agents." *Health Science Journal* 12 (5). https://doi.org/10.21767/1791-809x.1000596.

Gupta, Sajal, Meaghanne Caraballo, and Ashok Agarwal. 2018. "Total Antioxidant Capacity Measurement by Colorimetric Assay." In *Oxidants, Antioxidants, and Impact of the Oxidative Status in Male Reproduction*, 207–15. Elsevier. https://doi.org/10.1016/B978-0-12-812501-4.00019-5.

Hu, Bo, Shang Bing Wang, Kan Wang, Meng Zhang, and Shu Hong Yu. 2008. "Microwave-Assisted Rapid Facile 'Green' Synthesis of Uniform Silver Nanoparticles: Self-Assembly into Multilayered Films and Their Optical Properties."

Journal of Physical Chemistry C 112 (30): 11169–74. https://doi.org/10.1021/jp801267j.

Husen, Azamal, and Khwaja Salahuddin Siddiqi. 2014. "Phytosynthesis of Nanoparticles: Concept, Controversy and Application." *Nanoscale Research Letters* 9 (1): 1–24. https://doi.org/10.1186/1556-276X-9-229.

Ilyasov, Igor R., Vladimir L. Beloborodov, and Irina A. Selivanova. 2018. "Three ABTS•+ Radical Cation-Based Approaches for the Evaluation of Antioxidant Activity: Fast- and Slow-Reacting Antioxidant Behavior." *Chemical Papers* 72 (8): 1917–25. https://doi.org/10.1007/s11696-018-0415-9.

Ilyasov, Igor R., Vladimir L. Beloborodov, Irina A. Selivanova, and Roman P. Terekhov. 2020. "ABTS/PP Decolorization Assay of Antioxidant Capacity Reaction Pathways." *International Journal of Molecular Sciences*. MDPI AG. https://doi.org/10.3390/ijms21031131.

Iravani, S, H Korbekandi, S v Mirmohammadi, and B Zolfaghari. 2014. "Synthesis of Silver Nanoparticles: Chemical, Physical and Biological Methods." *Research in Pharmaceutical Sciences*. Vol. 9.

Jain, K. K. 2011. "Nanobiotechnology." *Comprehensive Biotechnology, Second Edition* 1: 599–614. https://doi.org/10.1016/B978-0-08-088504-9.00068-4.

Jain, Siddhant, and Mohan Singh Mehata. 2017. "Medicinal Plant Leaf Extract and Pure Flavonoid Mediated Green Synthesis of Silver Nanoparticles and Their Enhanced Antibacterial Property." *Scientific Reports* 7 (1): 1–13. https://doi.org/10.1038/s41598-017-15724-8.

Jalilian, Fereshteh, Azam Chahardoli, Komail Sadrjavadi, Ali Fattahi, and Yalda Shokoohinia. 2020. "Green Synthesized Silver Nanoparticle from *Allium ampeloprasum* Aqueous Extract: Characterization, Antioxidant Activities, Antibacterial and Cytotoxicity Effects." *Advanced Powder Technology* 31 (3): 1323–32. https://doi.org/10.1016/j.apt.2020.01.011.

Kaur, K, and Rebecca Thombre. 2021. *Chapter 1 Nanobiotechnology: Methods, Applications, and Future Prospects. Nanobiotechnology*. Elsevier Inc. https://doi.org/10.1016/B978-0-12-822878-4/00001-8.

Khan, Shakeel Ahmad, Sammia Shahid, and Chun Sing Lee. 2020. "Green Synthesis of Gold and Silver Nanoparticles Using Leaf Extract of *Clerodendrum inerme*; Characterization, Antimicrobial, and Antioxidant Activities." *Biomolecules* 10 (6). https://doi.org/10.3390/biom10060835.

Koduru, Janardhan Reddy, Suresh Kumar Kailasa, Jigna R. Bhamore, Ki Hyun Kim, Tanushree Dutta, and Kowsalya Vellingiri. 2018. "Phytochemical-Assisted Synthetic Approaches for Silver Nanoparticles Antimicrobial Applications: A Review." *Advances in Colloid and Interface Science*. Elsevier B. V. https://doi.org/10.1016/j.cis.2018.03.001.

Kut, Kacper, Bogumił Cieniek, Ireneusz Stefaniuk, Grzegorz Bartosz, and Izabela Sadowska-Bartosz. 2022. "A Modification of the ABTS• Decolorization Method and an Insight into Its Mechanism." *Processes* 10 (7). https://doi.org/10.3390/pr10071288.

Lomelí-Rosales, Diego Alberto, Adalberto Zamudio-Ojeda, Sara Angélica Cortes-Llamas, and Gilberto Velázquez-Juárez. 2019. "One-Step Synthesis of Gold and Silver Non-

Spherical Nanoparticles Mediated by Eosin Methylene Blue Agar." *Scientific Reports* 9 (1). https://doi.org/10.1038/s41598-019-55744-0.

Lomelí-Rosales, Diego Alberto, Adalberto Zamudio-Ojeda, Oscar Kevin Reyes-Maldonado, Morelia Eunice López-Reyes, Georgina Cristina Basulto-Padilla, Edgar José Lopez-Naranjo, Víctor Manuel Zuñiga-Mayo, and Gilberto Velázquez-Juárez. 2022. "Green Synthesis of Gold and Silver Nanoparticles Using Leaf Extract of *Capsicum chinense* Plant." *Molecules* 27 (5). https://doi.org/10.3390/molecules 27051692.

Mat Yusuf, Siti Nur Aishah, Che Nurul Azieyan Che Mood, Nor Hazwani Ahmad, Doblin Sandai, Chee Keong Lee, and Vuanghao Lim. 2020. "Optimization of Biogenic Synthesis of Silver Nanoparticles from Flavonoid-Rich *Clinacanthus nutans* Leaf and Stem Aqueous Extracts." *Royal Society Open Science* 7 (7). https://doi.org/10.1098/ rsos.200065.

Menezes, Bryan Brummelhaus de, Lucas Mironuk Frescura, Rafael Duarte, Marcos Antonio Villetti, and Marcelo Barcellos da Rosa. 2021a. "A Critical Examination of the DPPH Method: Mistakes and Inconsistencies in Stoichiometry and IC50 Determination by UV–Vis Spectroscopy." *Analytica Chimica Acta* 1157 (May). https://doi.org/10.1016/j.aca.2021.338398.

———. 2021b. "A Critical Examination of the DPPH Method: Mistakes and Inconsistencies in Stoichiometry and IC50 Determination by UV–Vis Spectroscopy." *Analytica Chimica Acta* 1157 (May). https://doi.org/10.1016/j.aca.2021.338398.

Mohammed, A. B. Abeer, Amr Mohamed, Noura El Ahmady El-Naggar, Hoda Mahrous, Ghada M. Nasr, Asmaa Abdella, Rasha H. Ahmed, Irmak, S., Elsayed, M. S. A., Selim, S., Elkelish, A., Alkhalifah, D. H. M. 2022. "Antioxidant and Antibacterial Activities of Silver Nanoparticles Biosynthesized by *Moringa oleifera* through Response Surface Methodology." *Journal of Nanomaterials* 2022 (Ccd). https://doi.org/10.1155/2022/9984308.

Mousavi-Khattat, Mohammad, Mehrnaz Keyhanfar, and Amir Razmjou. 2018. "A Comparative Study of Stability, Antioxidant, DNA Cleavage and Antibacterial Activities of Green and Chemically Synthesized Silver Nanoparticles." *Artificial Cells, Nanomedicine and Biotechnology* 46 (sup3): S1022–31. https://doi.org/ 10.1080/21691401.2018.1527346.

Nawaz, Haq, Momna Aslam, and Sidra Tul Muntaha. 2019. "Effect of Solvent Polarity and Extraction Method on Phytochemical Composition and Antioxidant Potential of Corn Silk." *Free Radicals and Antioxidants* 9 (1): 05–11. https://doi.org/10.5530/fra. 2019.1.2.

Nawaz, Haq, Muhammad Aslam Shad, Najiha Rehman, Hina Andaleeb, and Najeeb Ullah. 2020. "Effect of Solvent Polarity on Extraction Yield and Antioxidant Properties of Phytochemicals from Bean (*Phaseolus vulgaris*) Seeds." *Brazilian Journal of Pharmaceutical Sciences* 56. https://doi.org/10.1590/s2175-97902019000417129.

Nenadis, Nikolaos, and Maria Z Tsimidou. 2018. "*DPPH (2,2-Di(4-Tert-Octylphenyl)-1-Picrylhydrazyl) Radical Scavenging Mixed-Mode Colorimetric Assay(s).*"

Olszowy, Małgorzata. 2019. "What Is Responsible for Antioxidant Properties of Polyphenolic Compounds from Plants?" *Plant Physiology and Biochemistry*. Elsevier Masson SAS. https://doi.org/10.1016/j.plaphy.2019.09.039.

Olszowy, Małgorzata, and Andrzej L. Dawidowicz. 2018. "Is It Possible to Use the DPPH and ABTS Methods for Reliable Estimation of Antioxidant Power of Colored Compounds?" *Chemical Papers* 72 (2): 393–400. https://doi.org/10.1007/s11696-017-0288-3.

Oseguera-Galindo, David Omar, Eden Oceguera-Contreras, and Dario Pozas-Zepeda. 2020. "Silver Nanoparticles Synthesis Using Biomolecules of Habanero Pepper (*Capsicum chinense* Jacq.) as a Reducing Agent." *Journal of Nanophotonics* 14 (03): 036012. https://doi.org/10.1117/1.JNP.14.036012.

Phuyal, Nirmala, Pramod Kumar Jha, Pankaj Prasad Raturi, and Sangeeta Rajbhandary. 2020. "Total Phenolic, Flavonoid Contents, and Antioxidant Activities of Fruit, Seed, and Bark Extracts of *Zanthoxylum armatum* DC." *Scientific World Journal* 2020. https://doi.org/10.1155/2020/8780704.

Pradeep, Matam, Dariusz Kruszka, Piotr Kachlicki, Dibyendu Mondal, and Gregory Franklin. 2022a. "Uncovering the Phytochemical Basis and the Mechanism of Plant Extract-Mediated Eco-Friendly Synthesis of Silver Nanoparticles Using Ultra-Performance Liquid Chromatography Coupled with a Photodiode Array and High-Resolution Mass Spectrometry." *ACS Sustainable Chemistry and Engineering* 10 (1): 562–71. https://doi.org/10.1021/acssuschemeng.1c06960.

———. 2022b. "Uncovering the Phytochemical Basis and the Mechanism of Plant Extract-Mediated Eco-Friendly Synthesis of Silver Nanoparticles Using Ultra-Performance Liquid Chromatography Coupled with a Photodiode Array and High-Resolution Mass Spectrometry." *ACS Sustainable Chemistry and Engineering* 10 (1): 562–71. https://doi.org/10.1021/acssuschemeng.1c06960.

Rafique, Muhammad, Iqra Sadaf, M. Shahid Rafique, and M. Bilal Tahir. 2017. "A Review on Green Synthesis of Silver Nanoparticles and Their Applications." *Artificial Cells, Nanomedicine and Biotechnology* 45 (7): 1272–91. https://doi.org/10.1080/21691401.2016.1241792.

Rai, Mahendra, Alka Yadav, and Aniket Gade. 2008. "CRC 675 - Current Trends in Phytosynthesis of Metal Nanoparticles." *Critical Reviews in Biotechnology* 28 (4): 277–84. https://doi.org/10.1080/07388550802368903.

Shah, M. Z., Guan, Z.-H., Din, A. U., Ali, A., Rehman, A. U., Jan, K., Faisal, S., Saud, S., Adnan, M., Wahid, F., Alamri, S., Siddiqui, M. H., Ali, S., Nasim, W., Hammad, H. M., & Fahad, S. 2021. "Synthesis of Silver Nanoparticles Using *Plantago lanceolata* Extract and Assessing Their Antibacterial and Antioxidant Activities." *Scientific Reports* 11 (1): 1–14. https://doi.org/10.1038/s41598-021-00296-5.

Sinan, Kouadio Ibrahime, Mohamad Fawzi Mahomoodally, Ozan Emre Eyupoglu, Ouattara Katinan Etienne, Nabeelah Bibi Sadeer, Gunes Ak, Tapan Behl, and Gokhan Zengin. 2021. "HPLC-FRAP Methodology and Biological Activities of Different Stem Bark Extracts of *Cajanus cajan* (L.) Millsp." *Journal of Pharmaceutical and Biomedical Analysis* 192 (January). https://doi.org/10.1016/j.jpba.2020.113678.

Singh, Reetika, Christophe Hano, Gopal Nath, and Bechan Sharma. 2021. "Green Biosynthesis of Silver Nanoparticles Using Leaf Extract of *Carissa carandas* l. And Their Antioxidant and Antimicrobial Activity against Human Pathogenic Bacteria." *Biomolecules* 11 (2): 1–11. https://doi.org/10.3390/biom11020299.

Soural, I., P. Švestková, P. Híc, and J. Balík. 2022. "Different Values Obtained by the FRAP Method for the Determination of Slowly and Rapidly Reacting Phenols." *Acta Alimentaria* 51 (1): 84–92. https://doi.org/10.1556/066.2021.00168.

Sreelekha, E., Bini George, Aswathi Shyam, N. Sajina, and Beena Mathew. 2021. "A Comparative Study on the Synthesis, Characterization, and Antioxidant Activity of Green and Chemically Synthesized Silver Nanoparticles." *BioNanoScience* 11 (2): 489–96. https://doi.org/10.1007/s12668-021-00824-7.

Swilam, Noha, and Khaled A. Nematallah. 2020. "Polyphenols Profile of Pomegranate Leaves and Their Role in Green Synthesis of Silver Nanoparticles." *Scientific Reports* 10 (1): 1–11. https://doi.org/10.1038/s41598-020-71847-5.

Tatulian, Suren A. 2019. "FTIR Analysis of Proteins and Protein–Membrane Interactions." In *Methods in Molecular Biology*, 2003:281–325. Humana Press Inc. https://doi.org/10.1007/978-1-4939-9512-7_13.

Valgimigli, Luca, Andrea Baschieri, and Riccardo Amorati. 2018. "Antioxidant Activity of Nanomaterials." *Journal of Materials Chemistry B*. Royal Society of Chemistry. https://doi.org/10.1039/c8tb00107c.

Valsalam, Saritha, Paul Agastian, Mariadhas Valan Arasu, Naif Abdullah Al-Dhabi, Abdul Kareem Mohammed Ghilan, K. Kaviyarasu, Balasubramani Ravindran, Soon Woong Chang, and S. Arokiyaraj. 2019. "Rapid Biosynthesis and Characterization of Silver Nanoparticles from the Leaf Extract of *Tropaeolum majus* L. and It's Enhanced in-Vitro Antibacterial, Antifungal, Antioxidant and Anticancer Properties." *Journal of Photochemistry and Photobiology B: Biology* 191 (November 2018): 65–74. https://doi.org/10.1016/j.jphotobiol.2018.12.010.

Vuolo, Milena Morandi, Verena Silva Lima, and Mário Roberto Maróstica Junior. 2019. "Phenolic Compounds." In *Bioactive Compounds: Health Benefits and Potential Applications*, 33–50. Elsevier. https://doi.org/10.1016/B978-0-12-814774-0.00002-5.

Wang, Lu, Yanan Wu, Jia Xie, Sheng Wu, and Zhenqiang Wu. 2018. "Characterization, Antioxidant and Antimicrobial Activities of Green Synthesized Silver Nanoparticles from *Psidium guajava* L. Leaf Aqueous Extracts." *Materials Science and Engineering C* 86 (February): 1–8. https://doi.org/10.1016/j.msec.2018.01.003.

Wong, Pei Yun, and Seok Tyug Tan. 2020. "Comparison of Total Phenolic Content and Antioxidant Activities in Selected Coloured Plants." *British Food Journal* 122 (10): 3193–3201. https://doi.org/10.1108/BFJ-12-2019-0927.

Xiao, Fan, Tao Xu, Baiyi Lu, and Ruihai Liu. 2020. "Guidelines for Antioxidant Assays for Food Components." *Food Frontiers*. John Wiley and Sons Inc. https://doi.org/10.1002/fft2.10.

Yallappa, S., J. Manjanna, S. K. Peethambar, A. N. Rajeshwara, and N. D. Satyanarayan. 2013. "Green Synthesis of Silver Nanoparticles Using *Acacia farnesiana* (Sweet Acacia) Seed Extract Under Microwave Irradiation and Their Biological Assessment." *Journal of Cluster Science* 24 (4): 1081–92. https://doi.org/10.1007/s10876-013-0599-7.

Yousaf, Huma, Ansar Mehmood, Khawaja Shafique Ahmad, and Muhammad Raffi. 2020. "Green Synthesis of Silver Nanoparticles and Their Applications as an Alternative Antibacterial and Antioxidant Agents." *Materials Science and Engineering C* 112 (March): 110901. https://doi.org/10.1016/j.msec.2020.110901.

Zheng, Lin, Mouming Zhao, Chuqiao Xiao, Qiangzhong Zhao, and Guowan Su. 2016. "Practical Problems When Using ABTS Assay to Assess the Radical-Scavenging Activity of Peptides: Importance of Controlling Reaction PH and Time." *Food Chemistry* 192 (July): 288–94. https://doi.org/10.1016/j.foodchem.2015.07.015.

Chapter 4

Trends and Recent Advancements in the Application of Microbial Enzymes in Green Chemistry, Biotransformation and Sustainable Development

Amit Seth*
Department of Life Sciences (Botany), Manipur University, Canchipur, Imphal, India

Abstract

Developing useful products with biotechnological intervention has recently gained wide applicability and acceptability. Green chemistry and enzyme-mediated biotransformation are two aspects that have witnessed novel advancements having far-reaching implications. The amalgamation of applied microbiology, biotechnology, genetic engineering, immobilization, bioprocess development and downstream processing has inherently altered every facet of product synthesis. This chapter's main goal is to showcase microbial enzymes' role in green synthesis. The synthesis of useful products in various industries like medicine, chemical and agriculture is being pursued through enzyme-catalyzed processes. The main reason for this trend is the eco-friendly nature of enzymatic technologies required for attaining the twin goals of economic growth and sustainable development. Enzymes display multiple unique attributes making them highly indispensable for future industrial endeavors. Green chemistry aims to reduce the addition of toxic compounds into the environment, and one of the key technologies in realizing this objective is biocatalysis. With increasing emphasis on

* Corresponding Author's Email: amitsethshimla@gmail.com.

In: The Future of Green Synthesis
Editor: Jacques Kent
ISBN: 979-8-88697-816-2

scale-up and operational efficiency, key enzyme-based technologies have evolved and played a significant role in industrial production.

Keywords: enzyme catalysis, green chemistry, biocatalysis, biotransformation, bioproducts, sustainable development

Introduction

Microbial enzyme technology is a critical sector witnessing rapid advancements in recent times. The enzymes produced by microorganisms are quite varied and display exemplary potential. Microorganisms possess unique genetic diversity and thus can produce unique proteins. The microbial enzymes find wide application in varied fields like pharmaceutics, the chemical industry, and energy sectors (Blamey et al., 2017). These enzymes are a viable alternative to harsh chemical processes and thus have innumerable advantages. A concerted amalgamation of enzyme technology with advanced chemical processes stimulates rapid advancements in generating novel compounds. In earlier times, bacterial cells were the chosen source for commercial enzymes. However, with time, fungal cells were also explored to rapidly synthesize powerful enzymes (Ran et al., 2008).

Enzymes are vital for accelerating bioprocess operations. Every biological reaction can only be performed in the presence of the corresponding enzyme. Different types of enzymes have been successfully isolated from microorganisms and further characterized. These enzymes have been subjected to rigorous analytical procedures to make them more conducive for industrial applications. Although enzymes are ubiquitous in all organisms, microorganisms are the preferred source for their commercial production. The microscopic dimensions and small genome size make these microorganisms more amenable to scaling up operations. The global enzyme production market is expanding by leaps and bounds, thus signifying the usage of enzymes for diverse operations (Thapa et al., 2019). One key attribute that makes enzymes very successful in bioprocess operations is the property of enzyme specificity. Various traditional fermentation procedures known to man since time immemorial have been made plausible due to the activity of enzymes. However, their scientific aspects were revealed only with the development of the science of microbiology. Refinements in bioprocess technology and downstream processing added impetus to the field of enzyme technology. Agro-industry, pharma industry, leather manufacture, detergent production and textile industries increasingly use various enzymes for large-scale operations. Enzyme biocatalysis is steadily replacing traditional chemical-based reactions due to improved applicability and practicality associated with enzyme production and utilization (de Souza et al., 2016). Modern genetic engineering tools have further facilitated enzyme modification for improved production. However, some limitations associated with enzymes are the fragile nature of protein molecules and the requirement of specialized equipment for carrying out desired reactions. Enzyme production can only be carried out in well-equipped production centers requiring huge capital inputs. Thus, the hour needs to adopt low-cost production strategies for enzyme production. There is an urgent requirement to amalgamate enzyme technology with chemistry to develop novel chemoenzymatic routes for the optimal production of desired products. White biotechnology can help achieve the elusive goal of sustainable development by replacing toxic chemical reactions with eco-friendly enzyme synthesis (Arora, 2020).

Most commercial enzymes are hydrolytic. Various enzymes are metalloproteins requiring an inorganic cofactor for activation. The effectiveness of various chemical reactions can be markedly upgraded by infusing an enzyme concoction into the reaction mixture. Moreover, enzymatic reactions do not produce harmful by-products, making them highly amenable for food-grade operations. Enzyme specificity facilitates the fructification of hassle-free biochemical transformation under ambient conditions. Enzymes ensure the complete conversion of reactants into the desired products with negligible production of unwanted wastes, circumventing costly clean-up operations, a trait highly desirable for large-scale production. Thus, enzymatic product molecules are free from contaminating fragments, making them highly useful for therapeutic applications (Ramesh et al., 2020). Anti-microbial enzymes are used in the treatment of various infections. Most of the enzymes having biomedical relevance are recombinant origins. Microbes' genetic alteration has augmented these enzymes' robustness and usefulness. These modified enzymes can tolerate environmental fluctuations after administration in the patient's body and thus prove effective. The altered nature of therapeutic enzymes also enables them to dodge the host's immune response and achieve desired objectives without producing adverse side effects. The unique genetic makeup of microbes enables them to produce novel enzymes capable of synthesizing new metabolites. Extreme care should be taken while selecting microbial strains for enzyme production. The microbial enzyme source material ought to be non-virulent, accompanied by non-production of toxic bio compounds. The safety and risk assessment of the microorganism should be properly certified by authorized agencies for their effective usage (Gurung et al., 2013).

Although all microbes are potential sources of relevant enzymes, only a few select species or strains are suitable for commercial production. Most wild strains need to be genetically modified to make them commercially relevant. Genomic and proteomic advancements can make enzymes more relevant for green synthesis operations. Genetic improvement strategies can help regulate enzyme production, thus making it plausible to utilize substrate molecules effectively. New-age technologies, like metagenomics, can further deepen our knowledge of the unknown and explore our ecosystem's hitherto unexplored microbial wealth. A thorough understanding of microbes' growth and enzyme kinetics can further improve our understanding of enzyme mechanics. Bioreactor designing and statistical scale-up investigations can ensure greater usage of enzymes for amenable production of commodity chemicals. A multi-pronged approach will ensure the replacement of current chemical practices with vigorous biochemical reactions. Thus, greener and cleaner production targets can be made more achievable (Patel et al., 2017).

Green Synthesis and Whole-Cell Catalysis

Several biotransformation reactions are catalyzed by enantioselective enzymes, thus producing optically pure products. Whenever such reactions involve intracellular enzymes, whole microbial cells are used as a source of enzymes. These enantioselective enzymes are especially useful in the pharma sector for drug production. Whole-cell catalysis is economical as it helps sidestep costly protein purification procedures. Bacterial and yeast cells are added to reaction mixtures directly to achieve desired results. The efficiency of microbial cells can further be improved by subjecting these microbial strains to random and site-directed mutagenesis. The mutated strains can be directly utilized in biotransformation reactions (Luo et al., 2006). Recombinant strains of microorganisms are known for the hyper-expression of target protein molecules and the most common microbe used for heterologous expression in *E. coli*. Various strains of *E. coli* have been particularly developed for hosting foreign genes and facilitating their expression. Such recombinant strains often produce recombinant enzymes intracellularly and thus can be directly utilized as enzyme sources for accomplishing chemoenzymatic reactions. The utility can further be promoted by subjecting these microbial cells to various immobilization procedures to enhance their stability and reusability. The intracellular enzyme molecules are more stable than their extracellular counterparts. A recombinant cell's protective environment ensures the recombinant protein's long shelf life, thus enhancing its efficacy. Bioreactor designing can significantly enhance enzyme potency by undertaking fed-batch and continuous reactions. Both substrate and product tolerance can be enhanced by selective feeding strategies appropriate for higher production values (Peng et al., 2019).

Additionally, a consortium of microorganisms possessing divergent enzyme molecules can also be used for achieving the desired objectives and serve as alternatives for chemical transformations. Recent research investigations further indicate the increasing usage of fungal species in whole-cell catalysis. The eukaryotic cellular machinery of fungal cells further makes them more conducive to producing diverse metabolites that are useful for preparing pharmaceutical formulations. The fungal enzymatic products are bioactive and thus highly useful for human welfare. A multitude of chemical reactions can be facilitated by microbial cells as they comprise assorted enzymatic molecules. Many enzymes of fungal origin are also known to display appreciable stereo selectivity leading to the production of high-value products. The biotransformation spectrum is highly diverse for microbial enzymes, which is evident in the optimal production of divergent metabolites via a single microbial strain. By developing a better understanding of the cellular mechanisms, enzyme production and activity can be tightly regulated to ensure the production of novel biomedical derivatives (Choudhary et al., 2021). Whole-cell bio-catalysis is increasingly being used in the field of nanotechnology. Microorganisms have displayed a remarkable ability to produce nanomaterials via cost-effective alternate routes. The biological macromolecules housed in bacterial and fungal cells can facilitate the enzymatic conversion of precursor molecules into valuable nanomaterials. The reaction steps should be optimized minutely to ensure high production values. This enzymatic synthesis of nanomaterials facilitates the release of nano-products having desired morphological attributes. Chemically induced nanomaterial production requires harsh reaction conditions, whereas whole microbial cell-mediated biotransformation is effortlessly realizable in mild conditions (Deplanche et al., 2012).

Whole-cell catalysis ensures waste prevention, innocuous product formation, avoiding toxic solvent systems, recycling materials, decreasing reaction times, producing biodegradable alternatives and developing benign tactics. Whole-cell immobilization further leads to process refinement as it ensures catalytic reusability. The microbial biomass generated during large-scale reactions can easily be disposed of without adversely affecting our surrounding environment (Sheldon, 2016). Chemical transformations are usually carried out under extreme conditions of pH and temperature. This extreme environment is regarded as a general drawback. Fermentation procedures aided by microbial enzymes not only carry out these reactions under normal conditions but also help in achieving better conversion rates. In such microbe-assisted reactions, the co-cultivation of two or more species is increasingly becoming the new trend. Statistical optimization strategies lead to the creation of conducive conditions that support maximal microbial growth and favor high levels of expression of the target enzyme. These fermentation procedures usually last for 24-48 h, and occasionally lower yield of product formation is encountered. This shortcoming can be overcome by selecting robust strains of microbes that display high enzymatic activity. In this context, endophytic microorganisms associated with medicinal plants tend to display appreciable therapeutic potential. These endophytes have crucial enzymes that catalyze the synthesis of biological entities having high medicinal values (Jeandet et al., 2020).

An enzyme, substrate, and product are the three basic components of a biological process. Enzyme activity is profoundly influenced by both substrate and product. The substrate can lead to the saturation of active sites on enzyme molecules at higher concentrations, resulting in reduced activity. This condition is called substrate inhibition. It can be suitably overcome by limiting the substrate concentration in the reaction mixture below a certain threshold value by incorporating a fed-batch reactor into the operation (Raj et al., 2007b). Alternatively, product inhibition can be avoided by employing a continuous bioreaction process. As microorganisms contain various enzyme systems, a dual enzyme system may convert the desired product into a new intermediate. This problem is usually prevalent in whole-cell biotransformation, where complete microbial cells are used. Thus, it is essential to collect the target molecule well before being modified to a new compound in the reaction mixture. Another different approach is to induce the desired enzyme using specific inducers selectively. This selective induction process will ensure that only the main enzyme is activated and other residual enzymes remain inactive; thus, the accumulation of the target intermediate is ensured (Singh et al., 2021). In some instances, traditional mutagenic methods like physical or chemical mutagenesis have been successfully applied to the hyper-expression of intracellular enzymes. A comparative analysis of wild and mutant types can help identify specific base pair residues crucial for enhanced enzymatic activity. The mutated genes can subsequently be expressed in appropriate hosts like *E. coli* or *Pichia pastoris*. These recombinant cells on the intracellular production of the target enzyme can then directly be mixed with substrate molecules to achieve the required biotransformation (Pratush et al., 2017; Pratush et al., 2012).

Table 1. Role of whole cell catalysis in green synthesis

S. No.	Intracellular enzyme	Product synthesized	Advantage	Reference
1.	Alcohol dehudrogenase	Furoic acid	Efficient bioconversion of rice straw using recombinant whole cells of *E. coli*	Fotiadou et al., 2021
2.	Nitrile hydratase	Lactamide	Green synthesis of lactamide by a fed-batch method using hyper-induced bacterial cells	Singh et al., 2021
3.	Nitrile hydratase	Nicotinamide	Bioremediation of toxic nitrile compounds by a mutated enzyme expressed in *Pichia pastoris*	Pratush et al., 2017
4.	Nitrile hydratase	Butyramide	Green synthesis by whole cells of an actinomycetes bacteria	Raj et al., 2007a
5.	Nitrilase	Benzoic acid	Green synthesis by free and immobilized bacterial cells	Raj et al., 2007b

Chemical mutagenesis leads to the creation of a large mutant library that needs to be thoroughly screened for obtaining a hyperactive mutant type. Although the entire process is tedious, it remains a viable alternative. This traditional mutagenesis approach also suffers from an additional downside, which is the unstable nature of the mutated gene. Due to DNA repair mechanisms inherent in all organisms, the specific base pairs that have changed under the influence of the mutagen are reverted to their original form in subsequent generations. Nevertheless, a chemically mutagenized cell remains useful for whole-cell catalysis if it maintains its altered genetic structure (Pratush et al., 2010). The mechanism of enzyme activation and process development can be critically analyzed by isolating and purifying the target enzyme from the microbial cells. Enzyme kinetic analysis provides crucial insights into the enzyme's behavior in the presence of substrate molecules. New-age proteomic tools can help decipher the sequence alignment of constituent units and reveal the three-dimensional arrangement of amino acid residues. The operational parameters can be suitably modified considering this information and aid in bioreactor designing for the creation of an effectual milieu for the enzyme to function at its maximal potential (Pratush et al., 2013).

Occasionally, whole bacterial cells housing two or more enzyme systems may act concurrently during a single bioprocess. They tend to participate in a cascade reaction converting the substrate into two or more intermediate compounds. Accumulating more than one chemical component hampers downstream processing and hinders product refinement. Thus, undesirable by-product formation must be avoided during whole-cell biocatalysis. Closer analysis has revealed that the operational conditions for each enzyme are unique. Even though the enzymes originate in a single cell, they may get activated under diverse situations. A particular temperature or pH may activate one enzyme, whereas the other enzyme may not get stimulated in an identical setting. Thus, a tight regulation of bioreactor operational conditions can support the selective activation of the desired enzyme and lead to the accumulation of the desired transitional compound without the fear of its further alteration. This strategy is proved to be quite successful for nitrile hydratase and amidase enzymes reported from some actinomycete bacterial species (Raj et al., 2007a).

Green Synthesis and Enzyme Immobilization

Enzyme immobilization is an invaluable technique for achieving green chemistry goals. The economic costs of a bioprocess operation can be positively regulated by incorporating diverse enzyme immobilization methods. Large-scale production of commodity chemicals can be readily achieved using immobilized enzymes. Various materials ranging from high molecular weight polymers to magnetic nanomaterials are being extensively explored for successful enzyme attachment. The basic criterion for successful immobilization is the compatibility between the support structure and the target enzyme. Various materials are known to augment enzyme stability and efficiency. In many instances, the turnover number is markedly amplified for immobilized enzymes, coupled with the supplementary gain in enzyme reusability. The physical interaction of the enzyme and the support material can occur through many methods, including adsorption and entrapment. Steric hindrance is usually encountered during immobilization procedures; however, appropriate improvisation of the standard techniques can create better interaction and thus address these impediments appropriately. The immobilization matrices provide appropriate protection to the enzyme molecules and thus enhance their performance in challenging situations. The immobilized enzymes can display great potency even in unorthodox solvents, making them capable of catalyzing various green chemistry reactions (Fotiadou et al., 2021). New-age biomaterials have been developed and explored for diverse applications in recent years. The commercial applicability is determined by affordability and longer shelf life. Traditional matrices are being modified through enzymatic and non-enzymatic methods to fabricate novel derivatives that are more robust than their source materials. Various activating mediators or cross-linking agents alter enzyme support complexes' basic texture and chemical composition to augment their ability to attach enzyme molecules. Many immobilized enzyme structures in the form of beads, discs and gels have been developed based on the physical traits of the composites fused during the immobilization process. The uniform distribution of the enzyme molecules over the entire surface of the polymeric support structures is a prerequisite for successful enzyme immobilization. Alternatively, if the enzyme molecules are scattered unequally, then the desired biotransformation operations are significantly hampered. These limitations can be suitably overcome through appropriate monitoring of the fabrication process. The size of the interstitial spaces in the polymeric supports can be regulated by the appropriate use and selection of cross-linking activator molecules. Thus, enzyme loading and yield enhancement can be aptly realized. Any mismatch between the size of pores in the matrix and enzyme molecules can lead to enzyme leaching, thus hampering the entire bioprocess operation. In addition to size compatibility, another area of concern is the conformational state of the enzyme during the immobilized state. The support material should ensure the maintenance of the correct enzyme orientation to favor enzyme-substrate interaction for better product formation (Yassin et al., 2019).

Enzymes are protein molecules with a specific three-dimensional constituent amino acid arrangement. The amino acids are delicately arranged in a complex alignment that tends to get readily distorted in the event of little alterations in the surrounding environment, thus leading to loss of enzyme activity. However, most industrial reactions require robust protein molecules to function in harsh circumstances like organic solvents, pH, temperature fluctuations etc. Enzymes tend to behave differently in organic and aqueous solvents. Sometimes, the substrate molecules display a likeness to organic solvents, but the enzymes demonstrate better functionality in an aqueous environment. This differential affinity hampers enzyme-substrate interaction, thus reducing product yield (Molina-Gutiérrez et al., 2021).

Conversely, in immobilization, many chemical interactions between enzymes and support material uplift the enzyme's ability to tolerate fluctuating conditions. Agglomeration of enzyme molecules accommodates additional biocatalytic residues within a small area, and different active sites are made available for substrate interaction. Modern state-of-art spectroscopic technologies provide detailed insights into immobilized enzymes' structure and function. The development of novel support matrices facilitates rapid advancements in this field. Every aspect of the immobilization reaction is being carefully monitored, thus boosting enzyme efficacy. Comparative analysis of free and immobilized enzyme kinetics provides a clear roadmap for scale-up operations essentially required for commercial operations. The attached enzyme often exhibits limited flexibility due to additional bondages, leading to high K_m values as compared to the free enzyme (Rouhani et al., 2018).

Most enzymes obtained from bacteria and fungi are mesophilic and active at temperatures below 50°C. Enzyme immobilization enables biotransformation operations to be performed at elevated temperatures. Additional chemical interactions and extra protection of support material prevent enzyme denaturation and enhance bioconversion values. Successful enzyme immobilization should also ensure product synthesis in a minimal time. In situations where product formation is achieved over a long time, there is an enhanced plausibility that a reversible reaction may be triggered to convert the products back into the reactant. Thus, it becomes imperative that the desired goal of product formation is achieved in the shortest duration possible to avoid this limitation. The reversible reaction can only be avoided when there is a durable collaboration between the enzyme molecules and the support material and the enzyme can express its activity to its full capacity (Narwal et al., 2016). The applicability of enzyme immobilization technologies in critical pharma and food industry sectors requires using novel, non-toxic components that can replace certain harmful ingredients used in this process. Certain polymeric supports like chitosan have been proven highly beneficial for extracellular and intracellular enzymes. Whole-cell immobilization of bacterial cells enables high production yields and circumvents costly enzyme isolation and purification procedures. The spacer arm of the cross-linking reagent also facilitates the appearance of large pores on the matrix surface, thus enabling bulk immobilization of bacterial cells. Enzyme entrapment can be either partial or complete. Although partially entrapped cells have limited protection but usually encounter less stearic hindrance, thus having higher bioconversion potential. The bacterial entrapment method is particularly useful in bioremediation operations requiring the degradation of toxic compounds. The entrapped bacterial cell is sheltered from toxicogenic chemicals and can concurrently partake in the degradation process (Singh et al., 2020).

Table 2. Role of immobilized enzymes in green synthesis

S. No.	Immobilized enzyme	Immobilization Method	Application	Reference
1.	Lipase	Immobilization in Iron and Zinc magnetic nanoparticles	Green synthesis of lipophenols	Fotiadou et al., 2021
2.	Cellulase	Immobilization on carrageenan gel disks	Green synthesis of cellulose nanofibers	Yassin et al., 2019
3.	Lipase	Immobilization on Zinc and Manganese nanoparticles	Green synthesis of biodiesel	Molina-Gutiérrez et al., 2021
4.	Laccase	Immobilization on graphene oxide-based nanocomposites	Green synthesis of arylsulfonyl benzenediols	Rouhani et al., 2018
5.	Lipase	Immobilization on Silica	Green synthesis of isoamyl acetate	Narwal et al., 2016
6.	Nitrile hydratase	Immobilization on cross-linked chitosan	Green synthesis of lactamide	Singh et al., 2020
7.	L-methionine lyase	Immobilization on cellulose polymers	Green synthesis of volatile sulphur compounds	Abdelraof et al., 2020
8.	Nitrilase	Immobilization on cross-linked chitosan	Recyclable catalyst used in bioremediation of toxic nitrile compounds under ambient conditions	Jyoti et al., 2017

Enzyme immobilization is equally successful for enantioselective enzymes without hampering the stereoselective potency of the enzyme. Enzyme immobilization is also suitable for prolonged continuous bioreactor operations, although the reduced yield is witnessed in these protracted conditions. Immobilized enzymes can be used in packed bed reactors for optimal production and are also suitable for crucial chemoenzymatic transformations. Co-immobilizing enzymes can also execute multi-step reactions involving several enzymes in a single support matrix (Thompson et al., 2018). Partially purified enzymes have also displayed improvement in the immobilized state. The addition of the enzyme onto the carrier material is pH sensitive, and thus H^+ ion concentration should be tightly regulated to ensure maximal adherence. A change in pH alters the charge on the surface of enzyme molecules. Thus, the chemical interactions within the amino acid residues and amongst the enzyme and carrier molecules are tremendously affected (Abdelraof et al., 2020). Chitosan-like support materials function appropriately in an acidic environment. It has also been observed that substrate specificity is also dissimilar for free and immobilized enzymes. Substrates with the same functional groups but varying molecular weights interact differently with unbound and restrained enzymes. The diffusional barriers of the carrier material hamper the interaction of high molecular mass substrates with the enzyme molecules. Contrarywise, low molecular compounds owing to their relatively smaller size, can diffuse across the barricade and intermingle with enzyme residues resulting in better catalytic performance (Jyoti et al., 2017).

Green Synthesis and Enzyme Engineering

With the advent of genomics and proteomics, microbiology and enzyme technology science has witnessed a tremendous overhaul. Recombinant hosts and heterologous expression of the protein of interest have tremendously amplified the proficiency of industrial enzymes. Enzyme expression is being appropriately regulated to attain higher levels of product yield. Traditional random mutagenesis is now being replaced with targeted mutagenesis. A wide array of knowledge has been generated in the last two decades through refinements in sequencing technologies for both nucleic acids and proteins. Metabolomic profiling of useful microorganisms and their constituent proteins reveals new information that catapults the science of enzyme technology to a new level. Bioinformatic tools provide information about proteins' complex three-dimensional arrangement of amino acid residues. These *in silico* methodologies can predict enzyme behavior under diverse situations. Enzyme conversion rates improve as mutated proteins generate targeted products in record time. However, these techniques are high-end and thus require using costly and ultrapure biochemicals to perform these experiments in well-equipped laboratories (Cipolatti et al., 2019).

Recombinant DNA technology, on the one hand, is prompting a considerable improvement of prevailing bioprocesses while, on the other hand, stimulating the advancement of novel methodologies. The traditional method of extensive screening of microbes was quite cumbersome and time-consuming. Nowadays, the approach is to identify a few robust species of bacteria or fungi that demonstrate an intense growth and activity profile. These selected species are thoroughly investigated and can partake in the large-scale production of diverse compounds. Their cellular environment is modified to accommodate foreign genes and suitably assist their hyperexpression. A thorough investigation of diverse cellular pathways has revealed the nature and structure of key intermediates and the corresponding enzymes required for their synthesis. A considerable amount of these intermediates is vital for commercial activities, and thus their boosted accumulation in a cell is considered favorable. Many natural compounds are traditional products of plant and animal cells. Their production is limited, and insufficient amounts are produced in these eukaryotic cells. Recent advancements have enabled recombinant microbes to accept eukaryotic genes and ably assist in the expression of their products. The recombinant microbes can multiply very quickly and produce products rapidly.

Additionally, these microbes can be cultivated economically by utilizing low-cost carbon sources as media constituents (Maury et al., 2005). Novel fermentation pathways can be explored to replace the existing routes with the help of recombinant techniques. Transfer vectors capable of multiple expressions of two or more enzymes have been documented. This approach has helped in achieving productivity targets in record times. High-yield synthesis targets have been achieved with little or no unwanted chemicals produced, thereby achieving the objective of green synthesis. These novel routes have proved to be more successful compared to conventional pathways and are thus helping in overcoming the problems of conventional fermentation (Cui et al., 2021).

Enzyme engineering has made rapid strides in recent years due to the availability of powerful vector systems. They can carry transport genes and the main target gene into the recombinant strains. Protein expression levels are enhanced, and substrate utilization capabilities are upgraded. The substrate utilization spectrum of recombinant microbial strains is much better than wild ones. Recombinant enzymes possess unique active sites that accommodate diverse substrate residues, making high-value compound synthesis more plausible. Recombinant molecular technologies can help achieve the industry targets as the global demand for therapeutic products is ever-expanding (Rajesh et al., 2020).

Conclusion

Green chemistry stimulates interesting developments by exploring new avenues for synthesizing useful chemicals. The traditional chemical processes suffer from several voids and cannot obtain high yields of target molecules. The accumulation of undesired compounds also hampers product purification. Some of the ingredients added to the reaction are harsh and toxic. Given these drawbacks, other techniques are being explored. Enzyme-catalyzed reactions are performed under mild conditions and produce highly refined products. The side products are negligible and thus can be conveniently removed by downstream processing. Several microbial enzymes are also stereoselective and produce optically pure products. This attribute is especially relevant for pharma industries. The amalgamation of chemistry and enzyme technology has led to the formulation of novel chemoenzymatic pathways proving a pathbreaking alternative and signifying that interdisciplinary research and development is the need of the hour.

Enzyme discovery and evolution is a never-ending process. Enzyme refinement has ushered in a new era of eco-friendly, efficient technologies. It is our responsibility to downgrade the pollution levels on our planet. Dual application of enzymes is becoming the new norm wherein toxic compounds are being converted into useful compounds. Enzyme-based biomaterials are gaining traction in almost every sphere of human existence. The biotech industry is worth billions, radically stimulating rapid changes in all industries. Earlier the number of useful enzymes was very limited, but in the two decades, this number has swelled with the discovery of several novel enzyme systems. Enzymes are proving to be miracle molecules of nature capable of stimulating even the most complex reactions. Even though chemical sciences have displayed tremendous improvement recently, enzyme catalysis is leading from the front in promoting green chemistry.

Modern biotechnology, protein engineering, metagenomics and bioinformatics are unmasking the hitherto hidden facets of microbial enzymes. The mass cultivation of microorganisms in less time also leads to the rapid accumulation of enzyme molecules. The fermentation procedures are also undergoing remarkable changes due to novel bioreactor design and operation. Scale-up and commercial potential are the new buzzwords in this field. Enzyme reusability is economically viable that is reducing production costs. The human population is ever expanding with new demands, and enzyme-based green synthesis can meet the vital challenge of sustainable development.

Several industrial units are facing waste disposal issues. The cost inputs of setting up treatment plants are quite high.

Moreover, efficiency is also constrained; hence pollution levels continue to rise. Green technology aided by useful enzymes can meet these challenges head-on and contribute enormously to the problem of effluent treatment. There is an urgent need to invest heavily in promoting new research in enzyme technology to meet the urgent industry requirements. There are certain ethical issues involved in utilizing enzymes for therapeutic applications. There is also a danger of misuse of the power of recombinant DNA technology. A transparent and robust regulatory mechanism is essential for the optimal utilization of the power of biocatalysis in green synthesis. Moreover, many enzymes also play a vital role in food processing operations; consequently, an appropriate selection of non-toxicogenic microbes is essential. New-age cutting-edge technologies are a double-edged sword; this power must be utilized prudently to maximize the gains inherent in enzyme technology.

References

Abdelraof M, Ibrahim S, Selim M, Hasanin M. Immobilization of L-methionine γ-lyase on different cellulosic materials and its potential application in green-selective synthesis of volatile sulfur compounds. *Journal of Environmental Chemical Engineering*. (2020) 8(4): 103870. https://doi.org/10.1016/j.jece.2020.103870.

Arora R. Industrial potential of microbial enzymes. In *Microbial diversity, interventions and scope* (2020) (pp. 301-318). Springer, Singapore. https://doi.org/10.1007/978-981-15-4099-8_17.

Blamey J M, Fischer F, Meyer H P, Sarmiento F, Zinn M. Enzymatic biocatalysis in chemical transformations: a promising and emerging field in green chemistry practice. In *Biotechnology of microbial enzymes* (2017) (pp. 347-403). Academic Press. https://doi.org/10.1016/B978-0-12-803725-6.00014-5.

Choudhary M, Gupta S, Dhar M K, Kaul S. Endophytic fungi-mediated biocatalysis and biotransformations paving the way toward green chemistry. *Frontiers in Bioengineering and Biotechnology*. (2021) 9:419. https://doi.org/10.3389/fbioe.2021.664705.

Cipolatti E P, Pinto M C, Henriques R O, da Silva Pinto J C, de Castro A M, Freire D M, Manoel E A. Enzymes in green chemistry: the state of the art in chemical transformations. *Advances in Enzyme Technology*. (2019) 137-151. https://doi.org/10.1016/B978-0-444-64114-4.00005-4.

Cui R, Liu Z, Yu P, Zhou L, Zhou Z. Biosynthesis of l-alanine from cis-butenedioic anhydride catalyzed by a triple-enzyme cascade via a genetically modified strain. *Green Chemistry*. (2021) 23(18): 7290-8. https://doi.org/10.1039/D1GC02244J.

de Souza Vandenberghe L P, De Carvalho J C, Libardi N, Rodrigues C, Soccol C R. Microbial enzyme factories: current trends in production processes and commercial aspects. In *Agro-Industrial Wastes as Feedstock for Enzyme Production* (2016) (pp. 1-22). Academic Press. https://doi.org/10.1016/B978-0-12-802392-1.00001-0.

Deplanche K, Merroun M L, Casadesus M, Tran D T, Mikheenko I P, Bennett J A, Zhu J, Jones I P, Attard G A, Wood J, Selenska-Pobell S. Microbial synthesis of core/shell gold/palladium nanoparticles for applications in green chemistry. *Journal of the Royal Society Interface*. (2012) 9(72):1705-12. https://doi.org/10.1098/rsif.2012.0003.

Fotiadou R, Chatzikonstantinou A V, Hammami M A, Chalmpes N, Moschovas D, Spyrou K, Polydera A C, Avgeropoulos A, Gournis D, Stamatis H. Green synthesized magnetic nanoparticles as effective nanosupport for the immobilization of lipase: Application for the synthesis of lipophenols. *Nanomaterials*. (2021)11(2):458. https://doi.org/10.3390/nano11020458.

Gurung N, Ray S, Bose S, Rai V. A broader view: microbial enzymes and their relevance in industries, medicine, and beyond. *BioMed Research International*. (2013) Article ID 329121. https://doi.org/10.1155/2013/329121.

Jeandet P, Sobarzo-Sánchez E, Silva AS, Clément C, Nabavi SF, Battino M, Rasekhian M, Belwal T, Habtemariam S, Koffas M, Nabavi SM. Whole-cell biocatalytic, enzymatic and green chemistry methods for the production of resveratrol and its derivatives. *Biotechnology Advances*. 2020 39:107461. https://doi.org/10.1016/j.biotechadv.2019.107461.

Jyoti, Bhatia K, Chauhan K, Attri C, Seth A. Improving stability and reusability of *Rhodococcus pyridinivorans* NIT-36 nitrilase by whole cell immobilization using chitosan. *International Journal of Biological Macromolecules*. (2017) 103: 8-15. https://doi.org/10.1016/j.ijbiomac.2017.05.012.

Lou W Y, Zong M H, Smith T J. Use of ionic liquids to improve whole-cell biocatalytic asymmetric reduction of acetyltrimethylsilane for efficient synthesis of enantiopure (S)-1-trimethylsilylethanol. *Green Chemistry* (2006) 8(2):147-55. https://doi.org/10.1039/B514999A.

Maury, J., Asadollahi, M. A., Møller, K., Clark, A., Nielsen, J. Microbial Isoprenoid Production: An Example of Green Chemistry through Metabolic Engineering. In: Nielsen, J. (eds) Biotechnology for the Future. *Advances in Biochemical Engineering/Biotechnology*. (2005) vol 100. Springer, Berlin, Heidelberg. https://doi.org/10.1007/b136410.

Molina-Gutiérrez M, Alcaraz L, López F A, Rodríguez-Sánchez L, Martínez M J, Prieto A. Immobilized Forms of the *Ophiostoma piceae* Lipase for Green Synthesis of Biodiesel. Comparison with Eversa Transform 2.0 and Cal A. *Journal of Fungi*. (2021) 7(10): 822. https://doi.org/10.3390/jof7100822.

Narwal S K, Saun N K, Dogra P, Gupta R. Green synthesis of isoamyl acetate via silica immobilized novel thermophilic lipase from *Bacillus aerius*. *Russian Journal of Bioorganic Chemistry*. (2016) 42(1): 69-73. https://doi.org/10.1134/S1068162016010118.

Patel A K, Singhania R R, Pandey A. Production, purification, and application of microbial enzymes. In *Biotechnology of Microbial Enzymes* (2017) (pp. 13-41). Academic Press. https://doi.org/10.1016/B978-0-12-803725-6.00002-9.

Peng B, Ma C L, Zhang P Q, Wu C Q, Wang Z W, Li A T, He Y C, Yang B. An effective hybrid strategy for converting rice straw to furoic acid by tandem catalysis via Sn-sepiolite combined with recombinant *E. coli* whole cells harboring horse liver alcohol dehydrogenase. *Green Chemistry*. (2019) 21(21):5914-23. https://doi.org/10.1039/C9GC02499A.

Pratush A, Seth A, Bhalla T. Generation of mutant of *Rhodococcus rhodochrous* PA-34 through chemical mutagenesis for hyperproduction of nitrile hydratase. *Acta Microbiologica et Immunologica Hungarica*. (2010) 57(2):135-146. https://doi.org/10.1556/amicr.57.2010.2.6.

Pratush A, Seth A, Bhalla T C. Cloning, sequencing, and expression of nitrile hydratase gene of mutant 4D strain of *Rhodococcus rhodochrous* PA 34 in *E. coli*. *Applied Biochemistry and Biotechnology*. (2012) 168(3):465-486. https://doi.org/10.1007/s12010-012-9790-9.

Pratush A, Seth A, Bhalla T C. Expression of nitrile hydratase gene of mutant 4D strain of *Rhodococcus rhodochrous* PA 34 in *Pichia pastoris*. *Biocatalysis and Biotransformation*. (2017) 35(1):19-26. https://doi.org/10.1080/10242422.2016.1247831.

Pratush A, Seth A, Bhalla T C. Purification and characterization of nitrile hydratase of mutant 4D of *Rhodococcus rhodochrous* PA-34. 3 *Biotech*. (2013) 3(2):165-171. https://doi.org/10.1007/s13205-012-0081-5.

Raj J, Seth A, Prasad S, Bhalla T C. Bioconversion of butyronitrile to butyramide using whole cells of *Rhodococcus rhodochrous* PA-34. *Applied Microbiology and Biotechnology*. (2007a) 74(3):535-539. https://doi.org/10.1007/s00253-006-0693-y.

Raj J, Singh N, Prasad S, Seth A, Bhalla T. Bioconversion of benzonitrile to benzoic acid using free and agar entrapped cells of *Nocardia globerula* NHB-2. *Acta Microbiologica et Immunologica Hungarica*. (2007b) 54(1):79-88. https://doi.org/10.1556/amicr.54.2007.1.8.

Rajesh R O, Godan T K, Sindhu R, Pandey A, Binod P. Bioengineering advancements, innovations and challenges on green synthesis of 2, 5-furan dicarboxylic acid. *Bioengineered*. (2020) 11(1): 19-38. https://doi.org/10.1080/21655979.2019.1700093.

Ramesh A, Harani Devi P, Chattopadhyay S, Kavitha M. *Commercial applications of microbial enzymes*. In Microbial enzymes: roles and applications in industries (2020) (pp. 137-184). Springer, Singapore. https://doi.org/10.1007/s12033-019-00187-1.

Ran N, Zhao L, Chen Z, Tao J. Recent applications of biocatalysis in developing green chemistry for chemical synthesis at the industrial scale. *Green Chemistry*. (2008) 10(4):361-72. https://doi.org/10.1039/B716045C.

Rouhani S, Rostami A, Salimi A, Pourshiani O. Graphene oxide/$CuFe_2O_4$ nanocomposite as a novel scaffold for the immobilization of laccase and its application as a recyclable nanobiocatalyst for the green synthesis of arylsulfonyl benzenediols. *Biochemical Engineering Journal*. (2018) 133:1-11. https://doi.org/10.1016/j.bej.2018.01.004.

Sheldon R A. *Biocatalysis and green chemistry*. In Green Biocatalysis. (2016) (pp. 1-5). John Wiley & Sons. https://doi.org/10.1002/9781118828083.ch1.

Singh P, Kumari A, Attri C, Seth A. Efficient lactamide synthesis by fed-batch method using nitrile hydratase of *Rhodococcus pyridinivorans* NIT-36. *Journal of*

Microbiology, Biotechnology and Food Sciences. (2021) 567-572. https://doi.org/10.15414/jmbfs.2019/20.9.3.567-572.

Singh P, Kumari A, Chauhan K, Attri C, Seth A. Nitrile hydratase mediated green synthesis of lactamide by immobilizing *Rhodococcus pyridinivorans* NIT-36 cells on N, N′-Methylene bis-acrylamide activated chitosan. *International Journal of Biological Macromolecules*. (2020) 161: 168-176. https://doi.org/10.1016/j.ijbiomac.2020.06.004.

Thapa S, Li H, OHair J, Bhatti S, Chen F C, Nasr K A, Johnson T, Zhou S. Biochemical characteristics of microbial enzymes and their significance from industrial perspectives. *Molecular Biotechnology*. (2019) 61(8): 579-601. https://doi.org/10.1007/s12033-019-00187-1.

Thompson M P, Peñafiel I, Cosgrove S C, Turner N J. Biocatalysis using immobilized enzymes in continuous flow for the synthesis of fine chemicals. *Organic Process Research & Development*. 2018 Dec 10;23(1):9-18. https://doi.org/10.1039/C0SC00338G.

Yassin M A, Gad A A, Ghanem A F, Rehim M H. Green synthesis of cellulose nanofibers using immobilized cellulase. *Carbohydrate Polymers*. (2019) 205: 255-60. https://doi.org/10.1016/j.carbpol.2018.10.040.

Biographical Sketch

Amit Seth

Affiliation: Manipur University, India

Education: PhD

Business Address: Department of Life Sciences (Botany), Manipur University, Canchipur, Imphal Pincode – 795003 INDIA

Research and Professional Experience: 15 years teaching and research experience

Professional Appointments:

- Associate Professor, Department of Life Sciences (Botany), Manipur University, Imphal (2020-present)
- Associate Professor, Faculty of Applied Sciences and Biotechnology, Shoolini University, Solan (2019-2020)
- Assistant Professor, Faculty of Applied Sciences and Biotechnology, Shoolini University, Solan (2011-2019)

- Assistant Professor and Head of Department, Department of Botany, Government Post Graduate College, Karsog, District Mandi, (H.P.) (2010-2011)
- Two years teaching experience at Department of Biotechnology, Himachal Pradesh University, Shimla for M.Sc. Microbiology course (2008-2010)

Honors:

Name of Award	Awarding Agency	Year
Outstanding reviewer award	International Journal of Biological Macromolecules (Impact Factor 8.025), Elsevier	2018
Recognized reviewer award	International Journal of Biological Macromolecules (Impact Factor 8.025), Elsevier	2018
Recognized reviewer award	Process Biochemistry (Impact Factor 4.885), Elsevier	2019
S.S. Bir Gold Medal in Pteridology	Indian Fern Society	2018
CSIR-SRF	Council of Scientific and Industrial Research (CSIR)	2008
CSIR-UGC NET (For Lectureship)	UGC/CSIR	2003 and 2005
Certificate of Merit	Indian Fern Society	2014
International Travel Grant	DST	2017
Editorial Board Member	Vegetos (Springer Journal)	2022
Topic Editor	Frontiers in Bioengineering and Biotechnology (Impact Factor 6.064), Frontiers	2022

Publications from the Last 3 Years:

(1) Amit Seth, Ena Devi, Kiran Thakur, Chandrika Attri, Vijay Singh, Ayesha Bhandari, Mahavir Singh, M.K. Seth 2022. Himalayan Fern *Cheilanthes bicolor* Mediated Fabrication and Characterization of Iron Nanoparticles with Antimicrobial Potential. *Bio Nano Science* 12: 486–495. https://doi.org/10.1007/s12668-022-00969-z [Scopus listed].

(2) Ishita Guleria, Amita Kumari, Marie-Aleth Lacaille-Dubois, Adesh K. Saini, Vikas Kumar, Reena V. Saini, Uma Ranjan Lal, Naseem A. Gaur, Sonam Kumari, Amit Seth, Jyoti Dhatwalia, Shabnam Thakur and Sohan Lal 2022 In-vitro antimicrobial, antioxidant, anti-inflammatory,

and cytotoxic activities of *Populus ciliata* bark and leaves: A comparative study. *South African Journal of Botany* 148: 238-250. https://doi.org/ 10.1016/j.sajb.2022.04.040 [IF: 3.111].

(3) Surendirakumar, Kannaiah, Radha Raman Pandey, Thangavelu Muthukumar, Anbazhagan Sathiyaseelan, Surbala Loushambam, and Amit Seth 2022. Characterization and biological activities of melanin pigment from root endophytic fungus, *Phoma* sp. RDSE17. *Archives of Microbiology*, *204*(3), 171. https://doi.org/ 10.1007/s00203-022-02788-y [IF: 2.667].

(4) Kiran Thakur, Chandrika Attri, Amit Seth 2021. Nano-carrier based immobilization of enzymes for industrial application. *3 Biotech* 11(10), 1-12. https://doi.org/10. 1007/s13205-021-02953-y [IF: 2.893].

(5) Shailja Kumari, Amit Seth, Subhash Sharma, Chandrika Attri 2021. A holistic overview of different species of Potentilla a medicinally important plant along with their pharmaceutical significance: A review. *Journal of Herbal Medicine*. 29, 100460. https://doi.org/10.1016/j.hermed.2021.100460 [IF: 3.032].

(6) Paras Bohara, Kanika Dulta, Parveen Chauhan, Kiran Thakur, Amit Seth, P.K. Chauhan 2020. Impact of different solvents on phytoconstituents, antioxidant and FTIR analysis of *Diplazium esculentum* leaf extract from Himalayan region. *Plant Archives* 20(2): 6057-6063 [Scopus listed].

(7) Poonam Singh, Ansu Kumari, Kalpana Chauhan, Chandrika Attri, Amit Seth 2020. Nitrile hydratase mediated green synthesis of lactamide by immobilizing *Rhodococcus pyridinivorans* NIT-36 cells on N,N'-Methylene-bis-acrylamide activated chitosan. *International Journal of Biological Macromolecules* 161:168-176. https://doi.org/10.1016/j.ijbiomac.2020.06.004 [IF: 8.025].

(8) Shweta Chauhan, Varun Jaiswal, Chandrika Attri and Amit Seth 2020. Random mutagenesis of thermophilic xylanase for enhanced stability and efficiency validated through molecular docking. *Recent Patents on Biotechnology* 14: 5-15 http://dx.doi.org/10.2174/18722083136661907 19152056 [Scopus listed].

(9) Poonam Singh, Ansu Kumari, Chandrika Attri, Amit Seth 2019. Efficient lactamide synthesis by fed-batch method using nitrile hydratase of *Rhodococcus pyridinivorans* NIT-36. *Journal of Microbiology, Biotechnology and Food Sciences* 9(3): 567-572. https://doi.org/10. 15414/jmbfs.2019/20.9.3.567-572 [Scopus listed].

(10) Kavita Bhatia, Gorakh Mal, Rasbihari Bhar, Jyoti, Chandrika Attri and Amit Seth 2018 Purification and characterization of thermostable superoxide dismutase from *Anoxybacillus gonensis* KA 55 MTCC 12684. *International Journal of Biological Macromolecules* 117:1133–1139. https://doi.org/10.1016/j.ijbiomac.2018.06.031 [IF: 8.025].
(11) Jyoti, Kavita Bhatia, Kalpana Chauhan, Chandrika Attri and Amit Seth 2017 Improving stability and reusability of *Rhodococcus pyridinivorans* NIT-36 nitrilase by whole cell immobilization using chitosan. *International Journal of Biological Macromolecules* 103: 8–15. https://doi.org/10.1016/j.ijbiomac.2017.05.012 [IF: 8.025].

Chapter 5

Green Synthesis of Azole Derivatives as Antifungal Agents and Their Biomedical Applications

Trinath Biswal[1]
D. S. Ramakrishna[1]
Priyanka Behera[1]
and Sridhar Reddy Kothakapu[2]
[1]Department of Chemistry, Veer Surendra Sai University of Technology, Burla, Sambalpur, Odisha, India
[2]Department of 5-yr Integrated Chemistry, Palamuru University Mahabubnagar, Telangana, India

Abstract

Patients with immune compromise, such as those infected with HIV, organ transplant recipients, chemotherapeutic patients, and patients requiring long-term antimicrobial treatment, are increasingly difficult to treat. Hence, a comparatively new and remarkable class of antifungal drugs has to be developed in a cost-effective green synthesis pathway to solve this important medical problem. Since the last three years, from 2019 to 2022, the number of COVID-19 patients throughout the globe has been increasing, and its treatment is still a problem. No special drugs are produced for its treatment. In the last three decades, the diseases associated with invasive fungal infections have gradually increased throughout the world, and their proper treatment is a problem for medical science. In this regard, it was reported that some specific organic compounds, such as fluconazole, itraconazole, posaconazole, voriconazole, and ravuconazole, were found to be more influential in

In: The Future of Green Synthesis
Editor: Jacques Kent
ISBN: 979-8-88697-816-2

comparison with other antifungal drugs. The present article focuses on explaining the various methods of green synthetic approaches to Fluconazole, Itraconazole, Posaconazole, Voriconazole, and Ravuconazole using cost-effective and sustainable approaches.

Keywords: antifungal drugs, green synthesis, fungal infections, diseases

Introduction

During the last few decades, the increase in morbidity and mortality around the globe has been found to increase at a rapid rate because of fungal infections and some new, dangerous viruses [1–3]. There are three species of pathogens that are highly responsible for systemic fungal infections, such as Aspergillus species, Candida species, and Cryptococcus neoformans [4–7]. Hence, the treatment of viral and fungal infection diseases is highly challenging for human beings because the present traditional pathway of treatment for these kinds of diseases is not adequately effective to completely cure the diseases without any side effects [8–10]. In the present situation, of the four classes of important anti-fungal drugs such as azoles, polyenes, echinocandins, and antimetabolites, triazoles are found to be the most effective drug for the treatment of anti-fungal diseases [11]. The drugs related to azoles and azole-based compounds are capable of inhabiting the biosynthesis pathway of ergosterol, which is highly responsible for the formation of the fungal cell membrane [12–14]. The triazole drugs that are widely used for treating both superficial and systemic infections include fluconazole (FCZ), voriconazole (VCZ), itraconazole (ITZ), posaconazole (PSZ), and ravuconazole [15]. The structures of these drugs are shown in Figure 1.

The drugs containing triazole were found to be inevitable, which makes it highly necessary to recognize and synthesize novel medicines for anti-fungal management, which leads to the reduction of the various diseases caused by fungal infection.

- Fungal diseases.
- Number of hospital admissions
- Morbidity and mortality.

The conventional treatment of these kinds of diseases normally relies on identifying selective antifungal molecules in association with the therapeutic treatment. The major demerit of this kind of treatment process is as follows:

- time-consuming process that takes much longer than expected.
- An expensive and tedious path of treatment

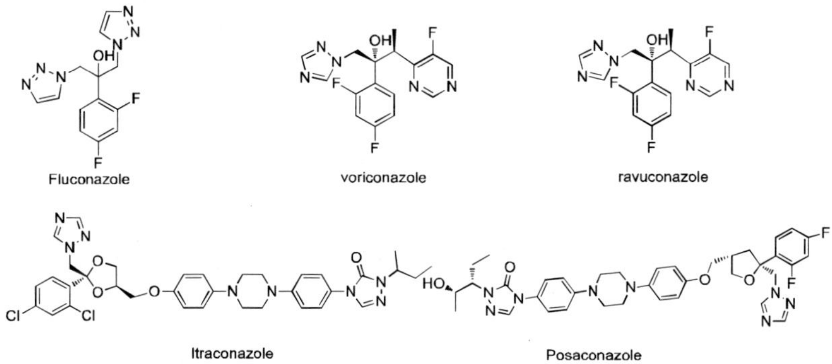

Figure 1. Structures of anti-fungal drugs.

Hence, we must develop and synthesize the drugs in a green and sustainable way, which can replace the traditional drugs and methods and provide a new, highly effective pathway for the treatment of diseases caused by fungal infection, becoming a new blessing to medical science. The first azole-based compounds with antifungal action were developed in 1944. After that, researchers were searching for drugs effective for antifungal diseases, chlormidazole was the first azole compound synthesized and made available for therapeutic use in 1958 [16]. Today, azole and its compounds got a lot of attention from researchers worldwide who are working to develop anti-fungal drugs. The most popular triazole antifungal medicines on the market include fluconazole, itraconazole, voriconazole, and posaconazole. Ravuconazole is a new triazole compound that is currently in phase 2 clinical trials and is found to be one of the most potent triazoles ever discovered. Researchers from around the world face challenges and opportunities for using this compound as an antifungal drug because the molecule of this compound contains two contiguous chiral centers, one of which is tertiary and the other quaternary [17].

Azole Drug: An Overview

Azole is an important class of antifungals in a group of medicines, that contains an azole ring in its structure and retards the growth of numerous fungi. These classes of drugs are classified into two groups, such as

- Two nitrogen atoms are present in the azole ring, which includes imidazoles (examples: clotrimazole, tioconazole, econazole, ketoconazole, and miconazole).
- Three nitrogen atoms are present in the azole ring, which includes triazoles (examples: fluconazole, voriconazole, itraconazole, and posaconazole).

Azole antifungal agents are massively used in the therapeutic options for the treatment of general fungal infections. It functions as an inhibitor towards the cytochrome P450-dependent enzyme lanosterol 14-alpha-demethylase that changes lanosterol to ergosterol, which is found to be the major main sterol in the cell membrane of fungi. The decrease in ergosterol causes the destruction of the cell membrane, which finally results in the death of the cell. The azole antifungal drugs can be successfully used for the treatment of in our skin and body. The most important development relating to the azole class of antifungal drugs has occurred during the 1990s, which has led to the treatment of numerous kinds of endemic and opportunistic fungal infections. Itraconazole and fluconazole were found to be comparatively safer than ketoconazole and amphotericin B. Despite many advances, it is highly difficult to treat some serious fungus-related diseases. The various anti-fungal drugs based on the azole group of compounds include fluconazole (diflucan (Pro)), itraconazole (Sporanox (Pro), itraconazole (Onmel (Pro)), clotrimazole (Mycelex), miconazole (Oravig (Pro)), ketoconazole (Nizoral (Pro)), itraconazole (Tolsura (Pro), voriconazole (Vfend (Pro), oteseconazole (Vivjoa (Pro), posaconazole (Noxafil (Pro), isavuconazonium (Cresemba (Pro) [18, 19].

Mechanism Action of the Azole Drugs

Azole antifungal agents usually retard the formation of ergosterol, a vital component of the fungal plasma membranes. This is owing to the inhabitation activity of the cytochrome P-450-dependent enzyme lanosterol demethylase.

This enzyme also plays a vital role in the synthesis of cholesterol in mammals. The presence of azole-based drugs in therapeutic concentrations and the efficacy of their antifungal activity are attributed to their higher affinity in the case of fungal P-450_{DM} as compared to the mammalian enzyme. The azole materials' long exposure to fungi leads to the accumulation of 140-methylated sterols and the depletion of ergosterol. The azole materials influence the functions of the ergosterol in fungal membranes, including the formation of chitin and nutrient transport. Ergosterol functions like a hormone in our body, effectively acting in the fugal cells that cause proliferation and growth. This function may also interrupt the depletion of ergosterol [20, 21].

Current Challenges of Using Azole Anti-Fungal Drugs

Numerous endemic and opportunistic fungal diseases undergo treatment due to the massive advance in azole-based antifungal agent-based materials during the 1990s. It was found that itraconazole and fluconazole are less toxic as compared with ketoconazole and amphotericin B. In spite of many developments in medical science, the proper treatment of fungal infections remains a challenging task for the global scientific community. Due to an increase in the resistance of antibiotic drugs, some serious fungal infections are highly difficult to treat, and the resistance to the commonly available drugs is developing. The effectiveness of azole materials is expected to increase if they are used in combination with some other suitable antifungal medicines having different mechanisms of action. The synthesis of triazoles of the second generation can also be effectively used in the treatment of endemic, opportunistic, and emerging fungal diseases. The antifungal agents of azole compounds used for treatment of antifungal diseases may be imidazoles or triazoles. Apart from ketoconazole, the application of imidazoles is highly restricted for the treatment of superficial mycoses, but the triazole class of azoles exhibits a wide range of applications for the treatment and cure of systemic and superficial fungal infections. Another important benefit of triazoles is their exceptional affinity for fungal infections rather than the mammalian cytochrome P-450 enzymes, which offers a better safety profile. Only the capsule formulation of itraconazole is available on the market, but recently the oral solution of itraconazole was approved [22, 23].

Some Important Class of Anti-Fungal Drugs Synthesized in Green Pathways

Some important classes of antifungal drugs effective for the treatment of various kinds of fungal diseases are as follows:

Fluconazole

Fluconazole is reported to be one of the most efficient first-generation antifungal medicines used for the treatment of various kinds of antifungal diseases. The fluconazole and its compound exhibit inhibitory activity on the cytochrome P450 enzyme 14α-demethylase and the inhibition of this enzyme prevents the synthesis of ergosterol, which is an important component of the fungal cell membrane. It is the most essential anti-fungal medicine approved by the WHO in 1990. The fungal infections that are efficiently cured by the use of this drug in proper doses include candidiasis, blastomycosis, coccidioidomycosis, croptococcosis, histoplasmosis, dermatophytosis, and pityriasis versicolor. Hence Fluconazole is highly important for the treatment of fungal and viral diseases [24].

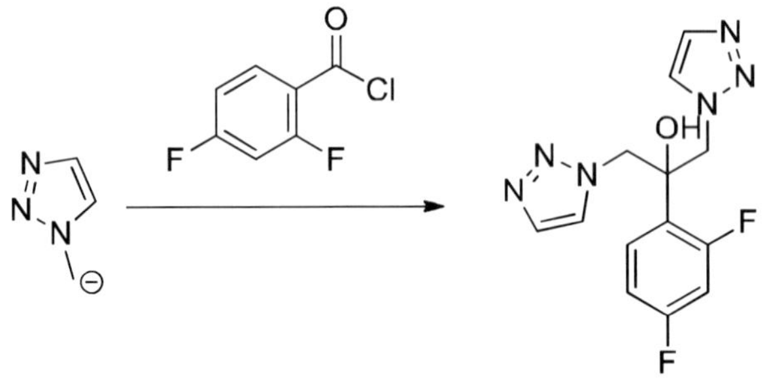

Scheme 1. Centro genesis route to Fluconazole.

The most remarkable synthetic strategy to synthesize fluconazole was reported by Judy Szeto et al., in the year 2019, by using a flow chemistry approach. Before this method of synthesis, some other important approaches were also reported in the literature, which required a greater number of steps, the purification of intermediates, longer reaction times, as well as the use of hazardous chemicals. Interestingly, the synthesis reported by Centro Genesis involved only two steps, but the yields were very poor [25] (Scheme 1).

Another two-step synthetic methodology was reported by Wu and co-workers, with the establishment of the following advantages:

- Simple commodity reagents such as α-chloroketone, triazole, and potassium carbonate
- Use of a class 3 solvent DMSO (Scheme 2). As per the green chemistry point of view, there are two major demerits:
  1) Purification of intermediates required much solvent.
  2) Formation of a greater number of by-products [26].

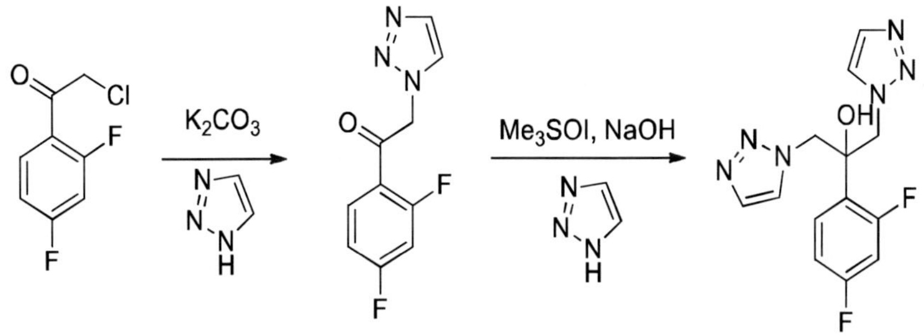

Scheme 2. Wu et.al. route towards Fluconazole.

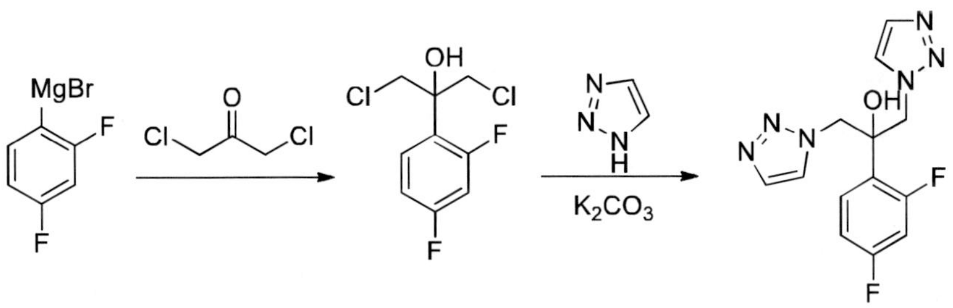

Scheme 3. Wang and co-workers' route to Fluconazole.

Wang and co-workers (2014) reported a new route to synthesize fluconazole that involved a Grignard reaction followed by a double alkylation reaction. To get better yields, suitable optimization conditions have to be implemented. But in this case, the reagent is highly hazardous to the environmental and biotic communities [27] (Scheme 3).

A modified version of a similar strategy developed by Wang and co-workers was reported by Jinan Luofeng Pharmaceutical with a change in the order of the alkylation and Grignard addition steps (Scheme 4). This route avoids the usage of more hazardous α, α, -dichloroacetone, and a better yield was obtained as compared to Wang and co-workers' route of synthesis [28].

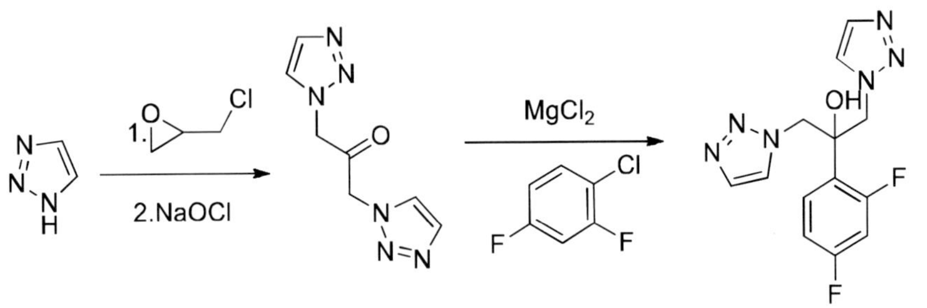

Scheme 4. Jinan Luofeng pharmaceutical technology company route to Fluconazole.

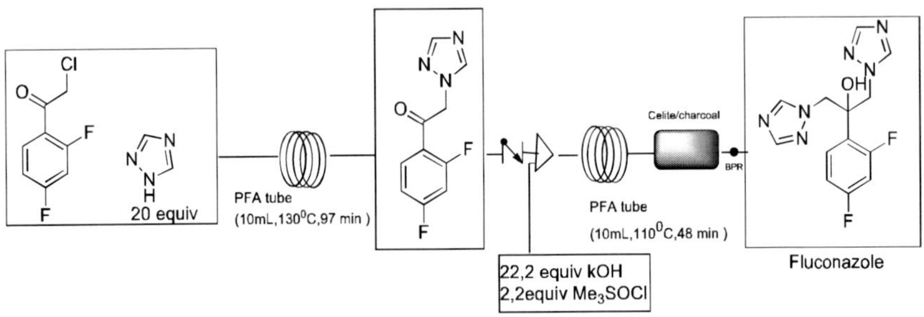

Scheme 5. Judy Szeto et.al. route to Fluconazole.

Finally, fluconazole was synthesized using a remarkable synthetic strategy with excellent yield by Judy Szeto et al. in 2019. This strategy has the following advantages:

- Multigrain is a compound that can be synthesized in a very short span of time, which satisfies the 6th green chemical principle.
- These reactions are simple enough to handle even with non-experts.
- As it doesn't require the purification of intermediates, it therefore satisfies the first green chemical principle [29].

Itraconazole

Itraconazole is a well-known triazole antifungal agent, which is basically used for the treatment of a broad spectrum of fungal infections, including aspergillosis, histoplasmosis, blastomycosis, coccidioidomycosis, and paracoccidioidomycosis. In addition to the fungal activity, it was also found that the inhibitory activity of itraconazole plays a significant role in the development of numerous cancer diseases. The sustainable synthetic strategy of synthesis starts with the initial route, which is a 10-step process with almost 6% overall yield by using the compound containing quaternary and tertiary chiral centres as the starting material [30]. Huge amounts of chemicals, such as organic solvents, were required to complete the synthesis in bulk amounts for commercialization and industrialization (Scheme 6) [31].

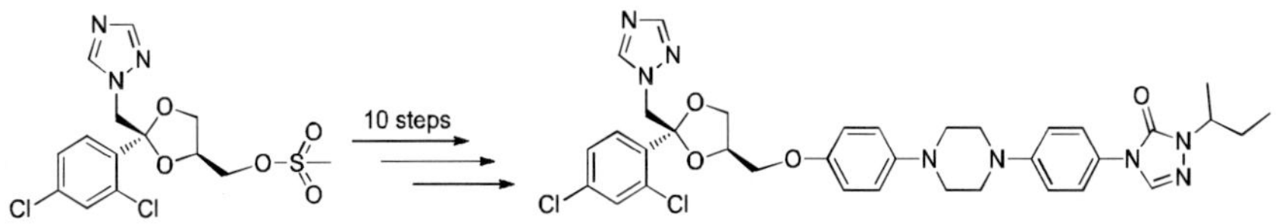

Scheme 6. Initial synthetic route to Itraconazole.

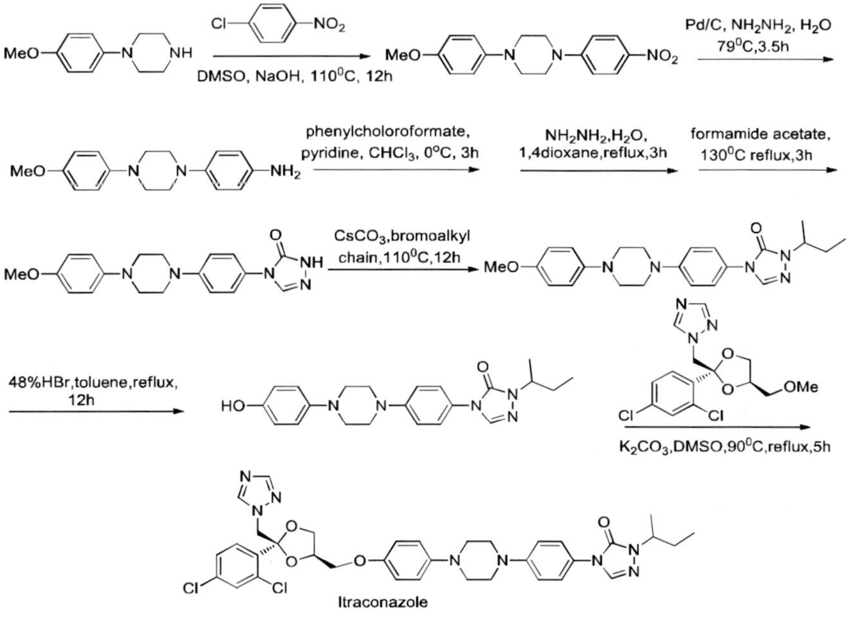

Scheme 7. Modified synthetic route to Itraconazole.

In continuation to the previous approaches, an eight-step green solvent-based process under highly complicated reaction conditions was established. Most of the time, DMSO is used, which is a class-3 organic solvent. The use of water as a green solvent provides an added advantage from a green chemistry point of view. This strategy was optimized for bulk-scale (100g) synthesis in the reactions. The execution of the first principle of green chemistry (prevention of waste formation instead of cleaning it up) in each step is the strategy that avoids the formation of side products in an appreciable quantity (Scheme 7) [32].

Posaconazole

- Posaconazole is a fluorinated anti-fungal and triazole-based drug. It was authorized by the US FDA in 2006 and is marketed under the brand names Noxafil and Posanol [28-31]. Posaconazole shows remarkable action against *Candida* species, *Cryptococcus* species, and endemic mycoses like *Coccidioides immitis*, *Histoplasma capsulatum*, and *Blastomyces dermatiditis*. In addition to its action against fungal infection, it is also effective in preventing transplantation complications.
- Since it was approved by the FDA, numerous synthetic strategies have been developed for the bulk preparation of this drug. The synthesis of any pharmacologically active ingredient and by-product residues are major concerns of this reaction approach in the case of Posaconazole preparations for commercialization [32–33]. Even though several reports are present for the preparation of Posaconazole, two important synthetic strategies are discussed here [33].

Kelly et al. reported a highly constrained route for the synthesis of Posaconazole in 2019. Multiple steps of the synthesis are the cause of the low yield of production (5%), which therefore makes industrialization not cost-effective. The major drawback of this method is the formation of comparatively more impurities and the requirement of expensive materials. The use of DCM in many steps is another significant drawback of this strategy (Scheme 8) [34].

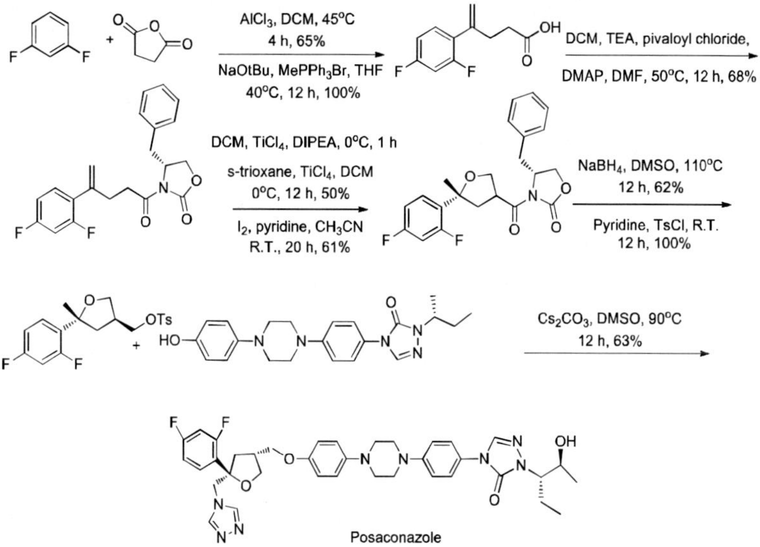

Scheme 8. Kelly et.al. synthetic route to Posaconazole.

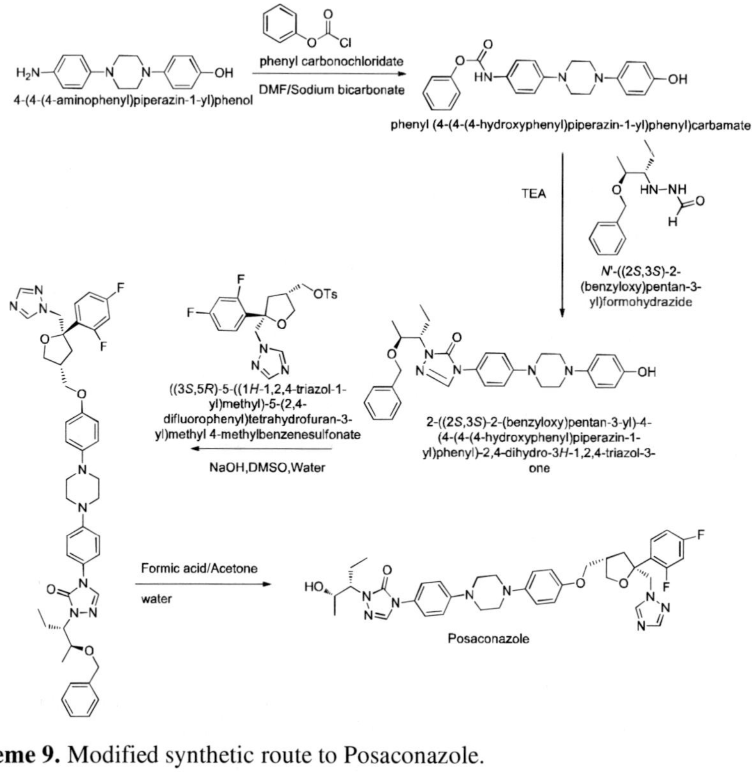

Scheme 9. Modified synthetic route to Posaconazole.

One of the modified methods towards the synthesis of Posaconazole has been reported by Krishnam Raju et al. in 2022. This route of synthesis reduces the formation of impurities as compared to all previous procedures adopted for the synthesis of this drug, which is an advantage with respect to green chemistry. In this approach, the percentage of yield was improved and the cost of the synthesis was reduced to an appreciable extent. One of the significant benefits of this method is the high purity of the product obtained, which is highly essential for the synthesis of any kind of drug. Since it is a cost-effective method, it is highly useful for bulk preparation of Posaconazole. In this process, palladium is never used during the debenzylation process. This reduces costs of preparation, reduces risks of health hazards, and eliminates the need for palladium to get finished product materials [35].

Voriconazole

Voriconazole is normally used for the treatment of patients affected by serious infections caused by yeast or fungi, such as fungal infection in the lungs (aspergillosis), fungal infection in the blood (candidemia), candida esophagitis (esophageal candidiasis), and some other fungal infections in the kidney, skin, stomach, wounds, bladder, etc. It is mainly used for the treatment of infections caused by fungi or yeast that the patient is unable to tolerate or in situations where no other medicine responds. It was observed that, with the passage of time, the resistance of any kind of fungus increases to drug Here, the resistance of Aspergillus to fluconazole is increased [36]. Today, a new triazole-based active substance has been developed and dubbed voriconazole, whose fluconazole structure has been modified to include the substitution of a 5-fluoropyrimidine ring for one of the triazole rings. It is a good antifungal medicine of the second generation, which is based on triazoles. Due to these structural alterations, voriconazole, also known as VCZ, is a triazolyl-butanol medicine that has two chiral centres and is effective against all kinds of Aspergillus species. Moreover, it was found that VCZ is more than 20 times more effective than FCZ against C. alb and C. gla. But organic chemists usually challenge voriconazole's stereochemical property of having two adjacent chiral centres [37]. The strong antifungal action of this drug needs a difluorobenzene moiety and a methyl group at the 3-position because it can resemble the 13-methyl unit of lanosterol, and the C3-methyl group present in it is also unquestionably important. A five-step synthesis method was reported for the synthesis of this drug, with the establishment of the following steps:

- Electrophilic substitution followed by nucleophilic substitution gives a ketone.
- Generation of two chiral centers with bromine reagent
- Final resolution step to get the desired compound. The use of DCM in two steps is a drawback to this route [38, 39].

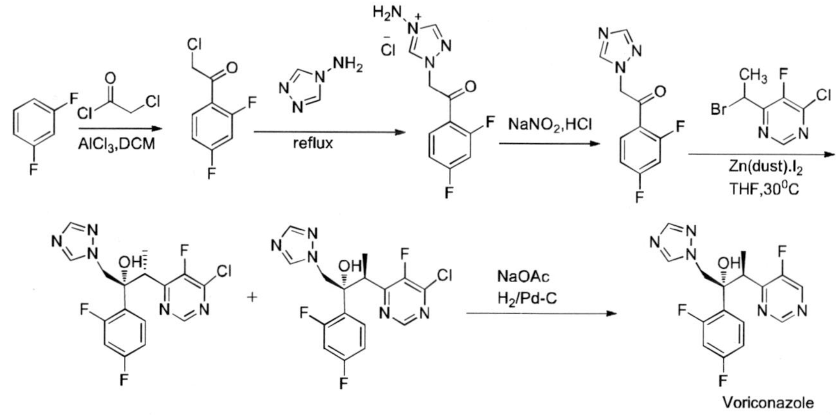

Scheme 10. A five step synthetic route to Voriconazole.

A modified similar strategy was established to voriconazole in two steps with excellent yield.

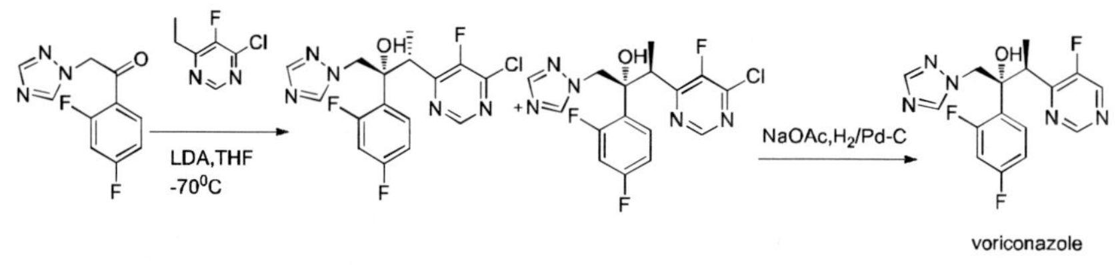

Scheme 11. A twostep synthetic route to Voriconazole.

Ravuconazole

Ravuconazole is a highly potent triazole antifungal drug, but its development was stopped in 2007. The spectrum of activity of this drug is like that of voriconazole, but with an increased half-life. But ravuconazole has specific action against species of Fusarium, such as Zygomycetes and Scedosporium. Ravuconazole and Posaconazole are triazole-based substances that prevent the

14α-demethylase or P450–CYP45-dependent enzymes, which are accountable for the ergosterol metabolism. Ergosterol is a vital component of the cell membrane of the parasite that cannot be found in the host. Ravuconazole is effective as a fungicide against moulds, with a bioavailability of 47% to 74% and a long half-life of around 100 hours. Chlormidazole was the first azole drug synthesized and made publicly available for medical use in 1958, just after the discovery of a new azole with antifungal activity in 1944. This is the most effectively used and under-researched class of antifungal agent. The well-known commercially available drugs of this kind include fluconazole, itraconazole, voriconazole, and posaconazole. Phase two clinical trials are currently being conducted on ravuconazole, which is a novel drug in this class. Every organic chemist will search for this compound, but it is difficult because it has two contiguous chiral centres, one of which is quaternary and the other is tertiary [40, 41].

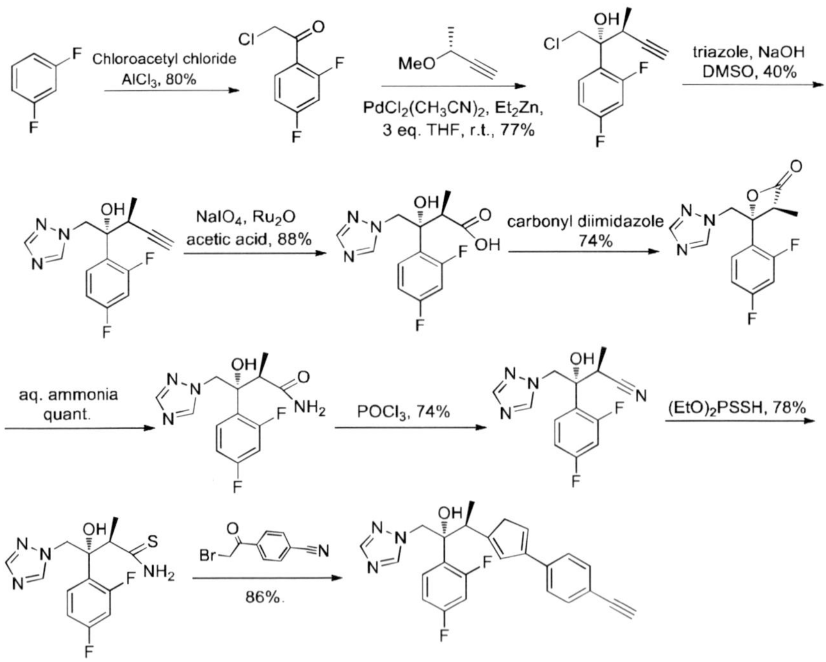

Scheme 12. Modified synthetic route to Ravuconazole.

Several efforts have been made to modify such kinds of molecules and develop the intermediate products to motivate the medical interest and prevent the difficulty of an anti-fungal drug. In this situation, M. Soukup et al.

synthesizes a a novel convergent ravuconazole drug for the treatment of fungal diseases [42]. A chiral propargylic mesylate and a zinc-allene derivative are combined in the crucial phase of this synthesis, which is based on a chiral allene addition reaction catalysed by palladium (Scheme 12). Lin Xu et al. described one of the simplified methods for synthesizing ravuconazole in 2009 by using homogeneous catalysis, but still there is a need for several modifications for the effective development of this drug for clinical purposes (Scheme 13) [43].

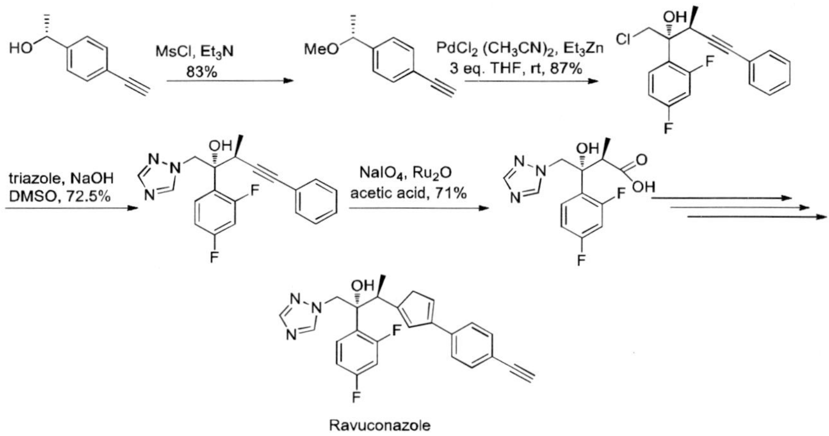

Scheme 13. Modified synthetic route to Ravuconazole.

Conclusion

Azole compounds have been used as drugs for the treatment of fungal infections for several decades, and they are one of the most used classes of antifungal agents. But researchers throughout the world are trying to develop new innovative techniques and strategies to get appropriate antifungal drugs with adequate safety, high antifungal potency, non-toxicity or less toxicity, high tolerability profiles, reduced drug–drug interactions, and high safety towards health. Although numerous azole-based substances have been designed and synthesized, because of their structural diversity, it is tough to simplify the nature and properties of the favourable substituents. Although the first synthesized azole substance is a popular pharmacophore pattern, the antifungal azole substances now exhibit excellent tolerance in their structural variation. Besides these, a recent approach is the inclusion of azole moieties in bioactive natural products, which can lead to the development of new substances with significant activity as antifungal agents, including against resistant strains. The optimization condition of synthesis approach was developed through a series of reactions for fluconazole, itraconazole, posaconazole, voriconazole, and ravuconazole, which function effectively as antifungals for the treatment of fungal diseases. In the case of fluconazole, a relatively lesser number of steps was found but for other antifungal azole-based substances, the number of steps was found to be higher for many of the synthetic strategies. But still, there is a great need to design and develop an ideal green synthetic route for the synthesis of this versatile anti-fungal material. Greener solvents were to be utilized in this regard by avoiding hazardous solvents such as DCM, THF, and DMF.

References

[1] Kidd, S. E., Chen, S. C. A., Meyer, W., Halliday, C. L. (2020). A New Age in Molecular Diagnostics for Invasive Fungal Disease: Are We Ready? *Front Microbiol*. 10, 2903- 2922. doi:10.3389/fmicb.2019.02903.

[2] Kainz, K., Bauer, M. A., Madeo, F., Carmona-Gutierrez D. (2020). Fungal infections in humans: the silent crisis. *MIC.,* 7(6), 143-145. doi:10.15698/mic2020. 06.718.

[3] Rodrigues, M. L., Nosanchuk, J. D. (2020). Fungal diseases as neglected pathogens: A wake-up call to public health officials. *Reynolds TB, ed. PLoSNegl Trop Dis.*14(2), 1-9. doi: 10.1371/journal.pntd.0007964.

[4] Bandalizadeh, Z., Javidnia, J., Hosseini, S. A., Moosazadeh, M., Amouei, A., Kermani, F., (2020). Cryptococcus and cryptococcosis in Iran during 1969–2019: A systematic review and meta-analysis. *Journal de Mycologie Médicale.*, 30(1), 1-33.

[5] Tamai, I. A., Pakbin, B., Fasaei, B. N. (2021). Genetic diversity and antifungal susceptibility of Candida albicans isolates from Iranian HIV-infected patients with oral candidiasis. *BMC Res Notes.*, 14(1). 93-99. doi:10.1186/s13104-021-05498-8.

[6] Arastehfar, A., Carvalho, A., Houbraken, J., Lombardi, L., Garcia-Rubio, R., Jenks, J. D. (2021.) Aspergillus fumigatusand aspergillosis: From basics to clinics. *Studies in Mycology*., 100(1), 100115-100115. doi: 10.1016/j.simyco.2021.100115.

[7] Nargesi, S., Abastabar, M., Valadan, R., Mayahi, S., Youn, J. H., Hedayati, M. T. (2021). Differentiation of Aspergillus flavus from Aspergillus oryzae targeting the cyp51A gene, *Pathogens* 10 (10). 1279-1283. https://doi.org/10.3390/pathogens101 01279.

[8] Alastruey-Izquierdo, A., Cadranel, J., Flick, H., Godet, C., Hennequin, C., Hoenigl, M. (2018). Treatment of Chronic Pulmonary Aspergillosis: Current Standards and Future Perspectives. *Respiration,* 96(2), 159-170. doi:10.1159/000489474.

[9] Slavin, M. A., Chen, Y. C., Cordonnier, C., Cornely, O. A., Cuenca-Estrella, M., Donnelly J. P. (2021). When to change treatment of acute invasive aspergillosis: an expert viewpoint. *Journal of Antimicrobial Chemotherapy*. 77(1),16-23.

[10] Xu, X., Naseri, A., Houbraken, J., Akbri, F., Wang, X., Zhao, R., (2021). Identification and in vitro antifungal susceptibility of causative agents of onychomycosis due to *Aspergillus* species in Mashhad, *Iran. Sci Rep*.,11(1). 6808-6815. doi:10.1038/s41598-021-86038-z.

[11] Howard, K. C., Dennis, E. K, Watt, D. S., Garneau-Tsodikova S. (2020). A comprehensive overview of the medicinal chemistry of antifungal drugs: perspectives and promise. *Chem Soc Rev*., 49(8), 2426-2480. doi:10.1039/c9cs00 556k.

[12] Irannejad, H., Emami, S., Mirzaei, H., Hashemi, S. M. (2020). In silico prediction of ATTAF-1 and ATTAF-2 selectivity towards human/fungal lanosterol 14α-demethylase using molecular dynamic simulation and docking approaches. *Informatics in Medicine Unlocked*., 20,1-9. doi:10.1016/j.imu.2020.100366.

[13] Cao, X., Sun, Z., Cao, Y., Wang, R., Cai, T., Chu, W. (2014). Design, Synthesis, and Structure–Activity Relationship Studies of Novel Fused Heterocycles-Linked Triazoles with Good Activity and Water Solubility. *J Med Chem*., 57(9), 3687-3706. doi:10.1021/jm4016284.

[14] Emami, S., Tavangar, P., Keighobadi, M. (2017). An overview of azoles targeting sterol 14α-demethylase for antileishmanial therapy. *European Journal of Medicinal Chemistry*., 135, 241-259. doi:10.1016/j.ejmech.2017.04.044.

[15] Nami, S., Aghebati-Maleki, A., Morovati, H., Aghebati-Maleki L. (2019) Current antifungal drugs and immunotherapeutic approaches as promising strategies to treatment of fungal diseases. *Biomedicine & amp; Pharmacotherapy*., 110, 857-868. doi: 10.1016/j.biopha.2018.12.009.

[16] Seehan, D. J., Hitchcock, C. A., Sibley, C. M. (1999). Current and emerging azole antifungal agents. *Clin. Microbiol. Rev*., 12 (1), 40–79. doi:10.1128/cmr.12.1.40.

[17] Becher, R., Wirsel, S. G. R. (2012). Fungal cytochrome P450 sterol 14α-demethylase (CYP51) and azole resistance in plant and human pathogens. *Appl Mirobiol Biotechnol*, 95(4), 825-840. doi:10.1007/s00253-012-4195-9.

[18] Fromtling, R. A. (1988). Overview of medically important antifungal azole derivatives. *Clinical microbiology reviews*, 1(2), 187-217. doi.org/10.1128/CMR. 1.2.187.

[19] Gavarkar, P. S., Adnaik, R. S., & Mohite, S. K. (2013). An overview of azole antifungals. International *Journal of Pharmaceutical Sciences and Research*, 4(11), 4083-4089.

[20] Georgopapadakou, N. H. (1998). Antifungals: mechanism of action and resistance, established and novel drugs. *Current opinion in microbiology*, 1(5), 547-557. https://doi.org/10.1016/S1369-5274(98)80087-8.

[21] Ghannoum, M. (2016). Azole resistance in dermatophytes: prevalence and mechanism of action. *Journal of the American Podiatric Medical Association, 106*(1), 79-86. doi: https://doi.org/10.7547/14-109.

[22] Srivastava, V., Singla, R. K., & Dubey, A. K. (2018). Emerging virulence, drug resistance and future anti-fungal drugs for Candida pathogens. *Current topics in*

medicinal chemistry, 18(9), 759-778. doi: https://doi.org/10.2174/1568026618666180528121707.

[23] Garg, A., Sharma, G. S., Goyal, A. K., Ghosh, G., Si, S. C., & Rath, G. (2020). Recent advances in topical carriers of anti-fungal agents. *Heliyon*, 6(8), 19-30.

[24] Abe, F., Usui, K., Hiraki, T. (2009). Fluconazole Modulates Membrane Rigidity, Heterogeneity, and Water Penetration into the Plasma Membrane in Saccharomyces cerevisiae. *Biochemistry*, 48(36), 8494-8504. doi:10.1021/bi900578y.

[25] . Korwar, S., Amir, S., Tosso, P. N., Desai, B. K., Kong, C. J., Fadnis, S., Telang, N. S., Ahmad, S., Roper, T. D., Gupton, B. F. (2017). The Application of a Continuous Grignard Reaction in the Preparation of Fluconazole. *Eur J Org Chem.*, 2017(44), 6495-6498. doi:10.1002/ejoc.201701002.

[26] Zhang, J., Liu, Z., Wu, H., & Cai, T. (2021). Effect of polymeric excipients on nucleation and crystal growth kinetics of amorphous fluconazole. *Biomaterials Science*, 9(12), 4308-4316.

[27] Wang, K., Jiang, S., Yang, Y., Fan, L., Su, F., & Ye, M. (2019). Synthesis and antifungal activity of carvacrol and thymol esters with heteroaromatic carboxylic acids. *Natural product research,* 33(13), 1924-1930.

[28] Wang, H. Q., Mou, S. B., Xiao, W., Zhou, H., Hou, X. D., Wang, S. J., ... & Xiang, Z. (2022). Structural Basis for the Friedel–Crafts Alkylation in Cylindrocyclophane Biosynthesis. *ACS Catalysis,* 12(3), 2108-2117.

[29] Collins, N., Stout, D., Lim, J. P., Malerich, J. P., White, J. D., Madrid, P. B., ... & Mallya, S. (2020). Fully automated chemical synthesis: toward the universal synthesizer. *Organic Process Research & Development,* 24(10), 2064-2077.

[30] Pace, J. R., DeBerardinis, A. M., Sail, V., Tacheva-Grigorova, S. K., Chan, K. A., Tran, R., Hadden, M. K. (2016). Repurposing the clinically efficacious antifungal agent itraconazole as an anticancer chemotherapeutic. *Journal of medicinal chemistry*, 59(8), 3635-3649. https://doi.org/10.1021/acs.jmedchem.5b01718.

[31] Liu, Y., Liu, Z., Cao, X., Liu, X., He, H., & Yang, Y. (2011). Design and synthesis of pyridine-substituted itraconazole analogues with improved antifungal activities, water solubility and bioavailability. *Bioorganic & medicinal chemistry letters*, 21(16), 4779-4783. https://doi.org/10.1016/j.bmcl.2011.06.062.

[32] Li, Y., Pasunooti, K. K., Peng, H., Li, R. J., Shi, W. Q., Liu, W., Liu, J. O. (2020). Design and synthesis of tetrazole-and pyridine-containing itraconazole analogy as potent angiogenesis inhibitors. *ACS Medicinal Chemistry Letters*, 11(6), 1111-1117.

[33] Rachwalski, E. J., Wieczorkiewicz, J. T., Scheetz, M. H. (2008). Posaconazole: An Oral Triazole with an Extended Spectrum of Activity. *Ann Pharmacother*, 42(10), 1429-1438. doi:10.1345/aph.1l005.

[34] Kesarwani, V., Kelly, H. G., Shankar, M., Robinson, K. J., Kent, S. J., Traven, A., & Corrie, S. R. (2019). Characterization of key bio–nano interactions between organosilica nanoparticles and *Candida albicans. ACS applied materials & interfaces*, 11(38), 34676-34687.

[35] Krishnam Raju, V. R., & Jha, A. (2022). An Improved Scalable Preparation of the Antifungal Posaconazole. *Organic Preparations and Procedures International*, 54(2), 191-195.

[36] Smith, J., Safdar, N., Knasinski, V., Simmons, W., Bhavnani, S. M., Ambrose, P. G., & Andes, D. (2006). Voriconazole therapeutic drug monitoring. *Antimicrobial agents and chemotherapy*, 50(4), 1570-1572. DOI: https://doi.org/10.1128/AAC.50.4.1570-1572.2006.

[37] Ghobadi, E., Saednia, S., & Emami, S. (2022). Synthetic approaches and structural diversity of triazolylbutanols derived from voriconazole in the antifungal drug development. *European Journal of Medicinal Chemistry*, 231, 114161. https://doi.org/10.1016/j.ejmech.2022.114161.

[38] Howard, K. C., Dennis, E. K., Watt, D. S., & Garneau-Tsodikova, S. (2020). A comprehensive overview of the medicinal chemistry of antifungal drugs: perspectives and promise. *Chemical Society Reviews*, 49(8), 2426-2480. https://doi.org/10.1021/cr040679f.

[39] Kronvall, G., & Karlsson, I. (2001). Fluconazole and voriconazole multidisk testing of Candida species for disk test calibration and MIC estimation. *Journal of Clinical Microbiology,* 39(4), 1422-1428. DOI: https://doi.org/10.1128/JCM.39.4.1422-1428.2001.

[40] Pasqualotto, A. C., Thiele, K. O., & Goldani, L. Z. (2010). Novel triazole antifungal drugs: focus on isavuconazole, ravuconazole and albaconazole. *Curr Opin Investig Drugs,* 11(2), 165-174.

[41] Yamaguchi, H. (2016). Potential of Ravuconazole and its Prodrugs as the New Oral Therapeutics for Onychomycosis. *Medical Mycology Journal*, 57(4), E93-E110. https://doi.org/10.3314/mmj.16-00006.

[42] Xu, L., Muller, M. R., Yu, X., & Zhu, B. Q. (2009). Improved chiral synthesis of ravuconazole. *Synthetic Communications,* 39(9), 1611-1625. https://doi.org/10.1080/00397910802563420.

[43] Liu, Y. L., & Lin, X. T. (2019). Recent advances in catalytic asymmetric synthesis of tertiary alcohols via nucleophilic addition to ketones. *Advanced Synthesis & Catalysis,* 361(5), 876-918. https://doi.org/10.1002/adsc.201801023.

Chapter 6

Synthetic Possibilities for Hydroxy-Methylenebisphosphonic Acids and Their Derivatives

Zsuzsanna Szalai*

Department of Organic Chemistry and Technology,
Faculty of Chemical Technology and Biotechnology,
Budapest University of Technology and Economics, Budapest, Hungary

Abstract

Many research groups around the world have been dealing with the synthesis of dronic acids and their derivatives for decades. These compounds are extremely effective in the treatment of osteoporosis and bone tumors. Their advantage is that they can be used in small doses, but due to their difficult preparation and complicated cleaning, their prices are very high [1, 2]. The industry's most commonly used solution for the synthesis of α-hydroxy-methylenebisphosphonates is the so-called "direct" synthesis, in which case the corresponding amino acids, heteroaryl acetic acids or derivatives are reacted with various P-reagents. The advantage of the method is that the reagents are easily accessible and relatively cheap, but their excessive use pollutes the environment. The disadvantages of this method are the extremely heterogenous reaction of the mixture, viscosity, difficulties in mixing and the released large amount of hydrochloric acid gas by the hydrolysis. These problems are manageable in laboratory environment, but it is difficult to perform these transformations in the industrial environment and to keep these reactions in safe conditions [3]. The "indirect" synthesis may also prepare the valuable drugs through α-oxophosphonates utilizing the Pudovik

* Corresponding Author's Email: szalai.zsuzsanna@edu.bme.hu.

In: The Future of Green Synthesis
Editor: Jacques Kent
ISBN: 979-8-88697-816-2

reaction [4-7]. There are other possibilities for the preparation of the active ingredients. Some of them are presented in this chapter.

Keywords: bisphosphonates, dronic acids, osteoporosis

Introduction

The human body contains 1000-2000 g of calcium, the majority (98%) is mostly found in bone tissue, in the form of hydroxyapatite [$Ca_{10}(PO_4)_6(OH)_2$]. It has an extremely important role because - among other things - it ensures the strength of the bone tissue. The amount of calcium entering the body is mainly influenced by the frequency and quality of food intake. The calcium concentration would fluctuate significantly if it only depended on this. If necessary, calcium could be mobilized from the bone or deposited there. This regulation is carried out by hormones, so changes in bone metabolism can fundamentally affect the functioning of the body.

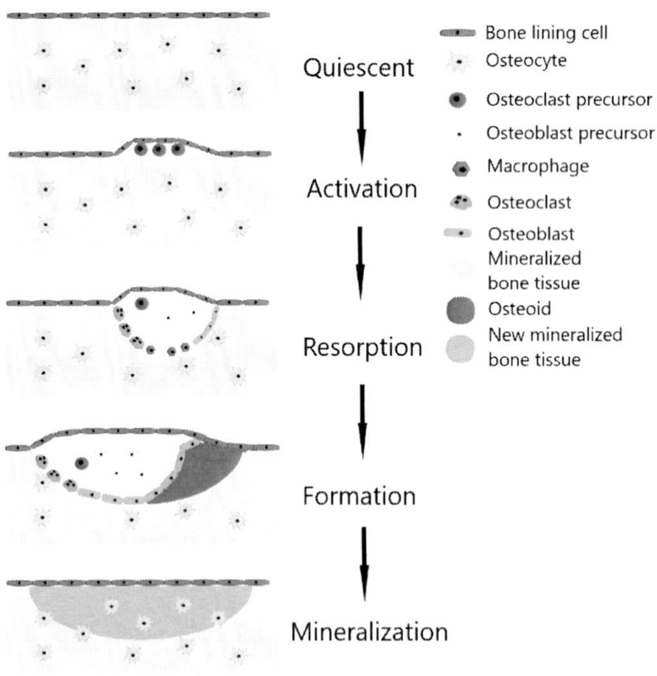

Figure 1. The cycle of bone remodeling [9, 10].

Bone tissue is constantly renewed during the life. In adulthood, remodeling occurs, during which osteoclasts appear in a specific location and begin bone resorption. After the formation of a small cavity, the monocyte/macrophages clean the surface, then the osteoblasts start building the new bone matrix, which will be later mineralized (Figure 1). In the process, bone resorption and bone formation must remain in balance, if this balance shifts in any direction, diseases develop. For instance, in the case of increased bone resorption, the consequence is osteoporosis [8, 9].

Table 1. Anti-bone resorption effect of some bisphosphonate agents and their use in the treatment of osteoporosis [8, 11]

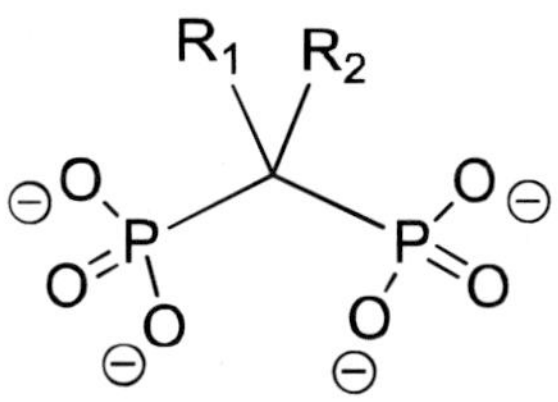

Bisphosphonates	R_1	R_2	Bone resorption inhibitory effect	Application	Active ingredient form
Etidronate	OH	CH_3	1x	per os	disodium salt
Clodronate	Cl	Cl	10x	per os, intravenous	disodium salt
Tiludronic acid	H	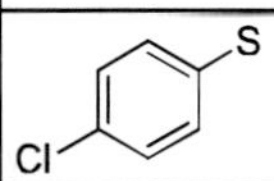	10x	intravenous	acid
Alendronate	OH	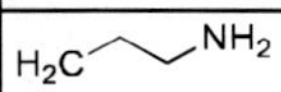	5.000x	per os	monosodium salt
Risedronate	OH	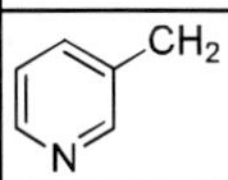	10.000x	per os	monosodium salt
Ibandronate	OH	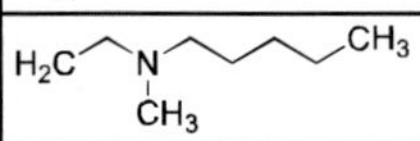	10.000x	per os, intravenous	monosodium salt
Zoledronic acid	OH	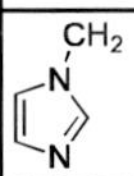	20.000x	intravenous	acid

It has long been known that polyphosphate, often used in washing powders, prevents the precipitation of calcium phosphates and carbonates. If a carbon atom takes place in this P–O–P bond instead of oxygen (P–C–P), then a new family of synthetic compounds is created, which *in vivo* inhibits the action of osteoclasts (bone resorption). These substances are called bisphosphonates. Many of them have already been commercially available as drugs: they differ chemically only in the substituent on the central carbon atom (Table 1) [8, 11].

Depending on the substituent on the central carbon atom, three groups can be distinguished: first, second and third generation compounds. Bisphosphonates in the first group do not contain an N atom in their side chain, such as etidronate, clodronate and tiludronic acid. The second generation dronates have an N-containing aliphatic substituent, for example, pamidronate, alendronate and ibandronate. Third generation bisphosphonates contain an N-heterocycle in their side chain, for example, risedronate and zoledronic acid (Table 1) [2, 12-14].

Etidronate was the first commercially available bisphosphonate. Unfortunately, the negative effect of the mentioned active ingredient is that it inhibits bone mineralization, which can lead to bone softening. For this reason, it was used only in small doses and intermittently. Newer derivatives have significantly better bone resorption-inhibiting effects and do not cause mineralization disorders. Clodronate and tiludronic acid are ten and alendronate five thousand times better in inhibiting bone resorption. Risedronate and ibandronate are ten thousand and zoledronic acid is twenty thousand times more effective than etidronate. Due to their effect, the amount of calcium flowing out of the bone decreases (Table 1) [8].

Preparation of Bisphosphonates by "Direct" Synthesis

The most commonly used solution in industrial environment for the synthesis of α-hydroxy-methylenebisphosphonates (**2**) is the so-called "direct" synthesis, in which case the corresponding amino acid, heteroaryl carboxylic acid or derivatives - which can be an acid chloride or anhydride – (**1**) is reacted with various P-reagents. This reagent is usually phosphorus trichloride, phosphorous acid, possibly phosphorus oxychloride or rarely, phosphorus pentachloride and phosphorus trioxide. The reaction temperature is 60-100°C. The reaction time varies widely, perhaps the longest is 24 hours (Scheme 1).

The desired bisphosphonate is obtained by hydrolysis and pH adjustment. The pure product is usually achieved by precipitation, digestion or recrystallization. Methanesulfonic acid (MSA), sulfolane, chlorobenzene, toluene, 1,4-dioxane, 1,2-dimethoxyethane, PEG-400 and n-octane are the most often used solvents. Also, the use of diethyl carbonate as a solvent can be mentioned as an environmentally friendly solution [3, 15-17].

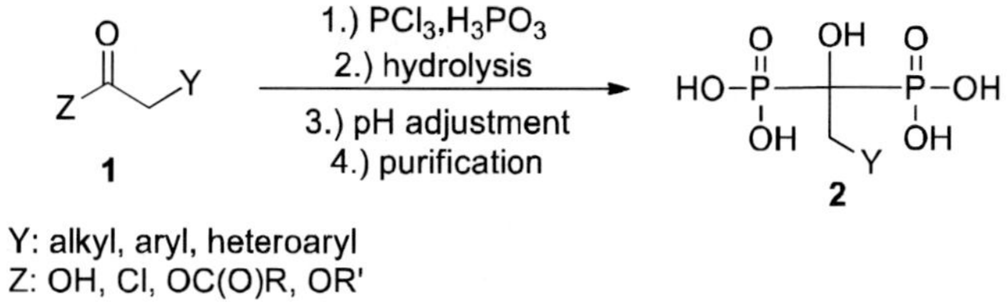

Scheme 1. "Direct" synthesis of dronic acids [3, 15-17].

Phosphorus reagents and solvents are often used in a large excess. These are easily accessible and relatively cheap. On the other hand, they pollute the environment, and the mixture is extremely heterogeneous, sticky, difficult to mix, and a large amount of hydrochloric acid gas is released during the hydrolysis. These problems are manageable in laboratory environment, but in the industry, it is already a problem to eliminate them. It is difficult to carry the reaction out under safe conditions. There are a wide range of yield data in the studies with only referring the raw product and without purity data which makes the comparison difficult. The high yields refer to acid-salt mixture which may also contain impurities [2, 18].

Synthesis of Zoledronic Acid under Different Conditions

Zoledronic acid is usually prepared by "direct" synthesis from imidazol-1-ylacetic acid (**3**). It belongs to the family of third generation dronats, in which the methylenebisphosphonate unit is linked to one of the nitrogens of the imidazole ring through a methylene group. The starting carboxylic acid is most often reacted with phosphorus trichloride and phosphorous acid. Several possible solvents are mentioned in the literature, their reactions carried out in methanesulfonic acid, sulfolane, or possibly ionic liquids (IL) mainly. It is common for phosphorous acid to be synthesized *in situ* by partial hydrolysis of 6 equivalents of phosphorus trichloride. The yield ranges from 24-86%, including pure dronic acid and Na salt of it. 30-40% of zoledronic acid (**4**) is obtained purely from the acid-salt mixture after recrystallization [2, 19].

McKenna et al., also investigated the synthesis of zoledronic acid under microwave (MW) conditions. Sulfolane was chosen as the solvent since its high polarity increases the efficiency of MW. With the help of MW irradiation, they were able to reduce the reaction time to less than 20 minutes, including hydrolysis. They could achieve good production, however, the authors did not provide purity data (Scheme 2) [20].

Two other research groups prepared the active ingredient (**4**) in n-octane and p-cresol. In the case of the former solvent, the acid-salt mixture was produced with 80% yield using 4.6 equivalents of phosphorus trichloride and the same amount of phosphorous acid. In the case of the latter, using 3.5 equivalents of phosphorus trichloride and 3 equivalents of phosphorous acid, the yield of the crude product was 65%. Purity data are not included in these cases either [21, 22].

The research group of Keglevich has been engaged the preparation and the study of bisphosphonates for more than a decade. Initially, third generation dronic acids including zoledronic acid (**4**) and risedronic acid (**6**) were synthesized (Scheme 3). The reaction parameters and yield data found in the studies showed high variability, the goal was to develop a reproducible synthesis, determining the optimal reaction conditions, molar ratios and reaction time [15].

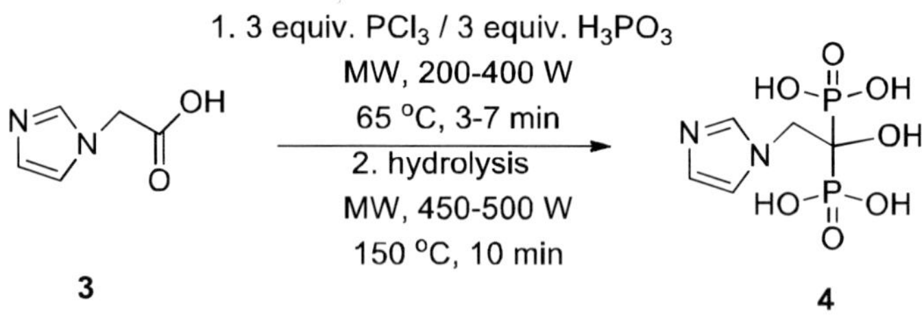

Scheme 2. Preparation of zoledronic acid under MW conditions [20].

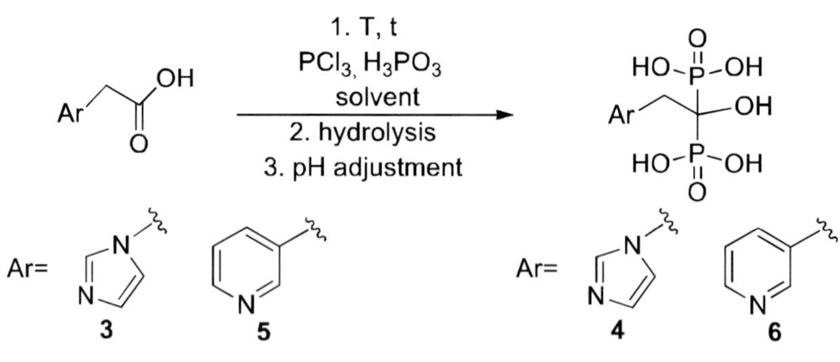

Scheme 3. Synthesis of zoledronic acid and risedronic acid [15].

The synthesis of zoledronic acid (**4**) was carried out in MSA at temperatures around 80°C, in the presence of phosphorus trichloride and phosphorous acid, starting from the corresponding carboxylic acid (**3**). After the reaction of the components, the mixture was hydrolyzed, and the pH of the reaction mixture was adjusted with sodium hydroxide. Dronic acid (**4**) and its sodium salt were obtained, as well as pure acid after recrystallization with hydrochloric acid. Purity was determined by potentiometric titration (Table 2) [2].

The results showed that the synthesis of zoledronic acid (**4**) does not require phosphorous acid, but phosphorus trichloride cannot be missed. In the presence of only phosphorous acid the reaction did not take place (Table 2, Entry 1). The reaction only phosphorus trichloride was carried out with, the yield is 53% and contained 99% zoledronic acid (**4**) (Table 2, Entry 2) [2].

Table 2. Molar ratios and the results obtained in the case of a reaction carried out in MSA [2]

Exp.	H_3PO_3 (Equiv.)	PCl_3 (Equiv.)	MSA (ml/g SM[a])	Reaction time (h)	Before recrystallization		After recrystallization	
					Yield[b] (%)	Acid content[b] (%)	Yield[c] (%)	Acid content[c] (%)
1	3.1	0	2.22	5	0	-	0	-
2.	0	3.1	3.33	3	71	26	53	99

[a] starting material (imidazol-1-ylacetic acid (**3**))
[b] for raw product
[c] based on potentiometric titration.

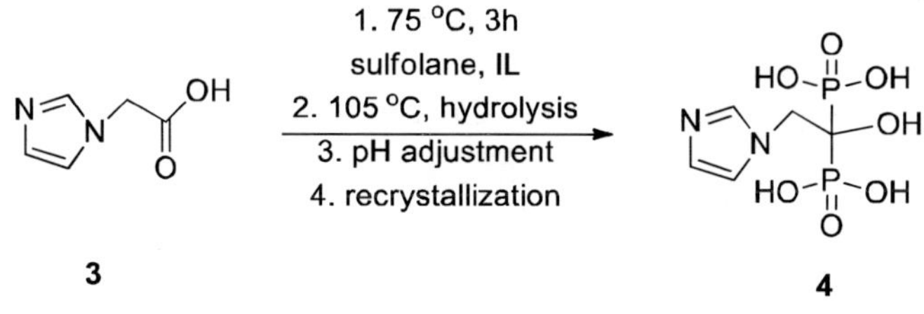

Scheme 4. Synthesis of zoledronic acid in sulfolane and ionic liquids [19].

Using 3.2 equivalents of phosphorus trichloride in sulfolane without the presence of phosphorous acid, the production was only 10% (Table 3, Entry 1), then in the presence of 3 equivalents of phosphorous acid, the yield of zoledronic acid (**4**) was 75%, the purity was 100% (Table 3, Entry 2) [19].

This method was extended to several other dronates (etidronate, phenidronate, pamidronate, alendronate, ibandronate), and in these cases, it was also concluded that phosphorous acid could be missed out if only using MSA as the solvent [3, 23, 24].

The synthesis of zoledronic acid (**4**) was carried out also in sulfolane, ionic liquids and mixtures of the two (Scheme 4) [19].

In the presence of an ionic liquid, the reaction proceeded similarly. 0.6 Equivalent of [bmim]BF_4 (1-Butyl-3-methylimidazolium tetrafluoroborate) was found to be optimal. The yield was 75% and the purity was 99% (Table 4, Entry 1). By practice when 1.2 equivalents of [bmim]BF_4 were added, the production decreased, it was only 49% (98% purity) (Table 4, Entry 2) [19].

In a mixture of sulfolane and ionic liquid, 0.6 equivalents of [bmim]BF_4 additives were added to the reaction mixture. In this case, the yield was 93%, with a purity of 99% (Table 4, Entry 3) [19].

Table 3. Molar ratios and results obtained in case of reaction in sulfolane [19]

Exp.	PCl_3 (Equiv.)	H_3PO_3 (Equiv.)	Yield (%)	Purity[a] (%)
1.	3.2	0	10	100
2.	3	3	75	100

[a] based on potentiometric titration.

Table 4. Preparation of zoledronic acid in the presence of an ionic liquid additive [19]

Exp.	PCl_3 (Equiv.)	H_3PO_3 (Equiv.)	[bmim][BF_4] Additive (Equiv.)	Yield (%)	Purity[a] (%)
1.	2 or 3	2	0.6[b]	75	99
2.	2 or 3	2	1.2[b]	49	98
3.	2	2	0.6[c]	93	99

[a] based on potentiometric titration
[b] in the presence of an ionic liquid additive, without solvent
[c] in the presence of an ionic liquid additive, in sulfolane.

When the syntheses were carried out in sulfolane without phosphorous acid, using only 3 equivalents of phosphorus trichloride no reaction occurred, and the product was formed only when both phosphorus reagents were present in the reaction mixture. On the other hand, in MSA phosphorous acid could be missed out. The conclusion is that MSA not only plays the role of solvent, but it is also present as a reaction partner [3].

The Mechanism of the Formation of Dronic Acids in Methanesulfonic Acid and Sulfolane Presented on the Example of Zoledronic Acid

MSA can react with phosphorus trichloride to produce an active phosphorus reagent and activate the starting carboxylic acid (**3**) to form a carboxylic acid-MSA mixed anhydride (**8**). The activated phosphorus reagent "attacks" the central carbon atom of this, and the carbonyl group of the carboxylic acid chloride (**7**) is formed from the carboxylic acid. An α-keto intermediate (**10**) is formed from the resulting compound (**9**). The "anhydride" reacts with the carbonyl group of the intermediate (**10**), and the resulting molecule (**11**) is hydrolyzed to obtain dronic acid (**4**) (Scheme 5) [3].

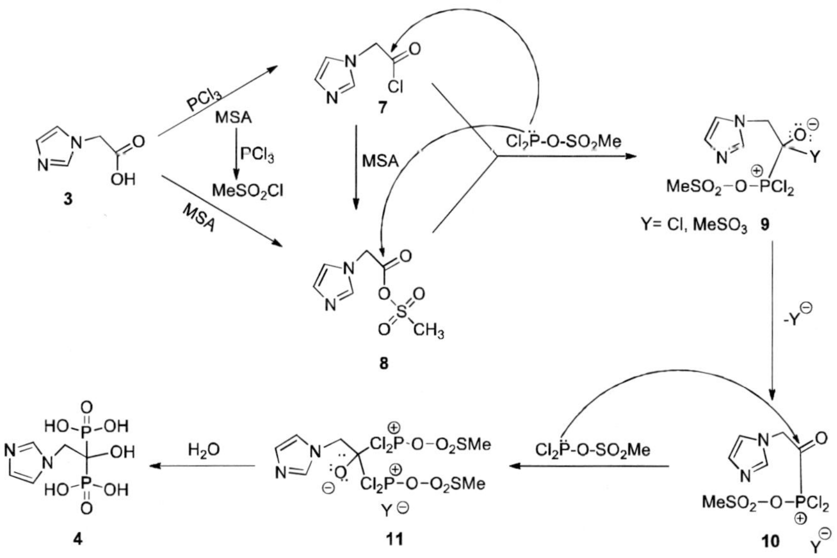

Scheme 5. The mechanism of zoledronic acid formation in MSA [3].

In sulfolane, one of the two molecules of phosphorus trichloride reacts with carboxylic acid (**3**) to form acid chloride (**7**), and the other two molecules react with phosphorous acid, thus forming a trimer condensate (**12**). This can react with the acid chloride (**7**) or the starting carboxylic acid (**3**), from which the resulting intermediate (**13**) is converted to ketophosphonic acid (**14**). Its carbonyl group is also reacted with an "anhydride," and by hydrolysis of the intermediate product (**15**) dronic acid (**4**) is obtained (Scheme 6) [3].

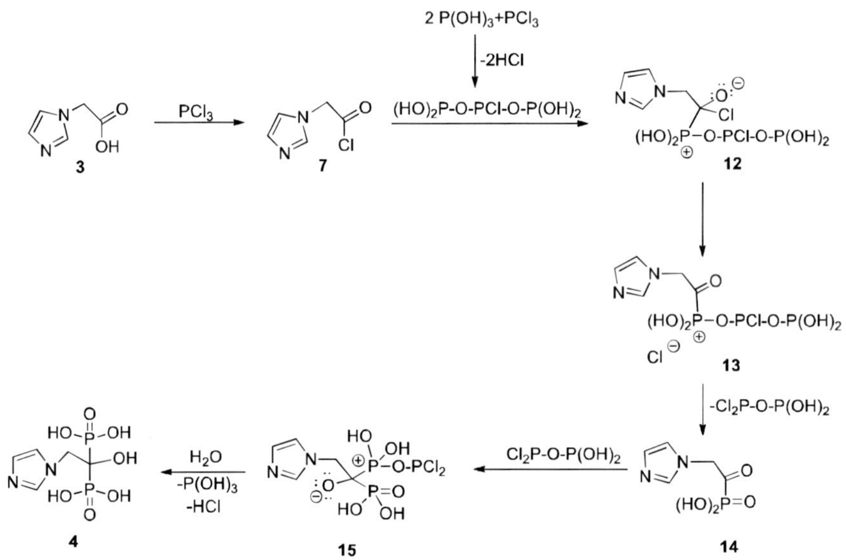

Scheme 6. The mechanism of zoledronic acid formation in sulfolane [3].

Preparation of Bisphosphonate Derivates by "Indirect" Synthesis

During the so-called "indirect" synthesis, an acid chloride (**17**) is prepared from the starting carboxylic acid (**16**), and then the resulting product (**17**) is converted into α-ketophosphonate (**18**) by adding trialkyl phosphite in the Michaelis-Arbuzov reaction. The desired product (**20**) obtained by an addition reaction with dialkyl phosphite and then by hydrolysis (Scheme 7) [5, 6, 25].

The Michaelis-Arbuzov reaction plays a key role in the synthesis of dialkyl or diaryl phosphonates and the formation of the P-C bond [26].

The Pudovik reaction is a method of forming a carbon-phosphorus bond, in which a reagent containing a labile phosphorus-hydrogen bond is added to the double bond of a C = X unsaturated compound (when C = X represents carbonyl or imine groups). In organophosphorus chemistry, the aza-Pudovik reaction is used to produce α-amino-methylphosphonates. In the presence of a base catalyst, the dialkyl phosphite phosphorus-hydrogen bond is added to the carbon-nitrogen double bond of the imine (hydrophosphonylation reaction). The reaction is closely related to the three-component Kabachnik-Fields reaction, in which we can talk about the condensation reaction of an amine, a phosphite and an oxo compound [27-30].

In 1956 McConnel and Coover established that the reaction of diethyl acetylphosphonate (**21a**) with dialkyl phosphites obtains α-substituted hydroxy-bisphosphonates (**22a**) in the presence of base catalysts. Acetylphosphonate (**21a**) was dripped into an equimolar mixture of diethyl phosphite and diethylamine (DEA) under solvent-free conditions. The product was characterized based on boiling point, refractive index, elemental analysis and infrared spectral data (Scheme 8) [31].

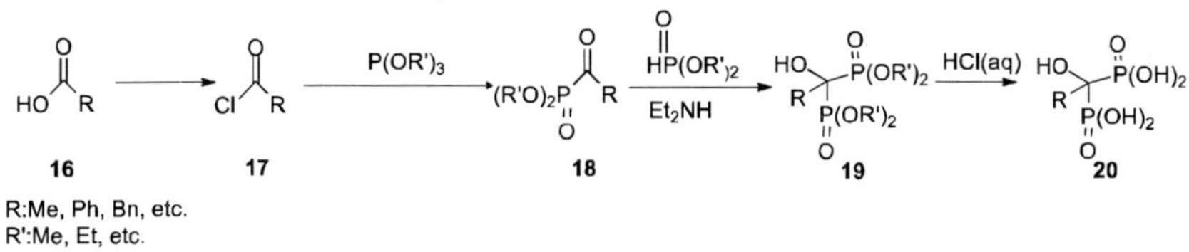

Scheme 7. The "indirect" synthesis of α-hydroxy-methylenebisphosphonic acid [5, 6, 25].

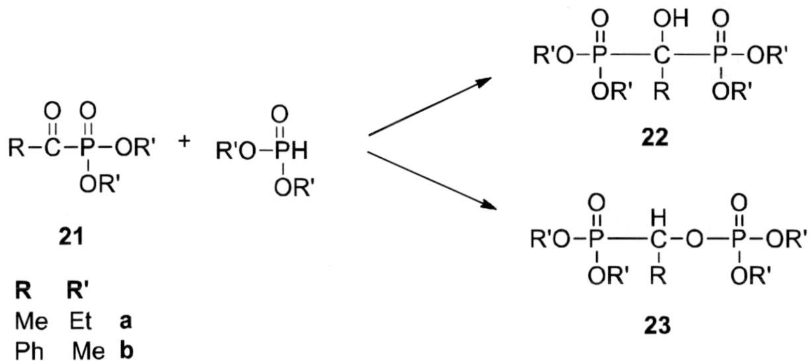

Scheme 8. The reaction of α-oxophosphonates and dialkyl phosphites [5, 31, 32].

A few years later (1962), Fitch and Moedritzer repeated these reactions and experienced that the corresponding phosphonate-phosphate (**23a**) was formed by rearrangement in the reaction. In the ^{31}P NMR spectrum, there were two signals at δ_P values of -1.3 and 20.2 [32].

In that case, if the reaction mixture was not heated above 80°C, the corresponding bisphosphonate (**22a**) was formed. There was one signal in the ^{31}P NMR spectrum (δ_P 20.8) [32].

Nicholson and Vaughn gave a detailed description of the preparation of the bisphosphonate (**22b**) by the reaction of dimethyl benzoylphosphonate (**21b**) and dimethyl phosphite in the presence of 0.05 equivalent of dibutylamine (DBA) (Scheme 8) [5].

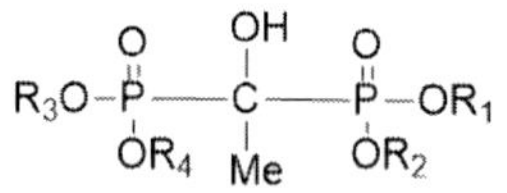

R_1	R_2	R_3	R_4	Yield (%)
Me	Me	Me	Me	89
Et	Et	Et	Et	92
^{i}Pr	^{i}Pr	^{i}Pr	^{i}Pr	81
Me	Me	Me	Ph	48
Et	Et	Me	Me	38
Ph	Ph	Me	Me	55

Figure 2. Some of the α-hydroxy-methylenebisphosphonates synthesized by Vepsäläinen's research group [33].

Decades later, the research group of Vepsäläinen and Turhanen prepared symmetric and asymmetric tetraalkyl α-hydroxy-methylenebisphosphonates based on Nicholson's method. The products were received with varying yields (16-92%), but no rearrangement was mentioned (Figure 2). Some tri-, di- and monoester derivatives were synthetized too [33, 34].

Hammerschmidt et al., studied the stereochemical course of the α-hydroxyphosphonate-phosphate rearrangement and determined that the configuration is conserved during the transformation [35, 36].

An Example for Pudovik Reactions under MW Conditions

Keglevich et al., was able to selectively produce both the diethyl acetyl-phosphonate-diethyl-phosphite adduct and the corresponding rearranged phosphonate-phosphate under solvent-free MW conditions [6, 37].

Among many others, the reaction of dimethyl (phenylacetyl)phosphonate (**24A**) and dimethyl phosphite was investigated. The ketophosphonate is in equilibrium with the corresponding enolphosphonate form (**24B**) (Scheme 9) [37].

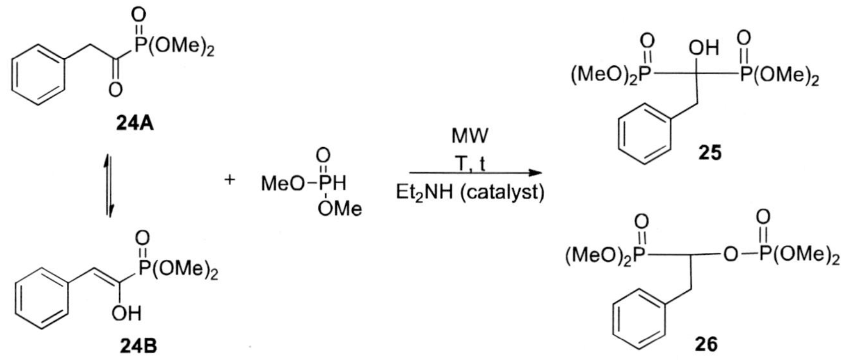

Scheme 9. The reaction of dimethyl (phenylacetyl)phosphonate and dimethyl phosphite under MW conditions [37].

Table 5. The reaction of dimethyl (phenylacetyl)phosphonate and dimethyl phosphite [37]

Exp.	T (°C)	t (min)	DEA (%)	Product composition (%) [a]		Yield (%)
				25	26	
1.	80	20	20	59	41	52 (26)
2.	120	10	20	0	100	

[a] Based on the relative ^{31}P NMR intensities.

Using 20% of DEA, the phosphonate-phosphate (**26**) also was formed, which was obtained from the product mixture with a yield of 52% (Table 5, Entry 1). At 120°C, only the formation of the rearranged product was observed (Table 5, Entry 2) [37].

Study of the Formation of Pudovik-Adducts and the Rearranged Products, Fine-Tuning of the Reaction Conditions

In the research group of Keglevich, we recently investigated the Pudovik reactions of α-oxophosphonates with dialkyl phosphites and diphenylphosphine oxide, and determined the conditions of the rearrangement process [7].

Among other things we studied the reaction of diethyl acetylphosphonate (**27a**) and diethyl benzoylphosphonate (**27b**) with dimethyl phosphite (Scheme 10) [7].

Table 6. The reaction conditions of diethyl α-oxophosphonates and dimethyl phosphite [7]

Exp.	Z	Catalyst (%)	Product composition (%) [a]			Yield (%)
			28	29	30	
1.	Me	DEA (5)	94	4	2	
2.	Me	DEA (40)	3	76	21	76 (29a+30a)
3.	Me	DBA (5)	100	0	0	87 (28a)
4.	Ph	DEA (5)	-	81	19	80 (29b+30b)

[a] Based on the relative ^{31}P NMR intensities.
[b] It rearranges while standing.

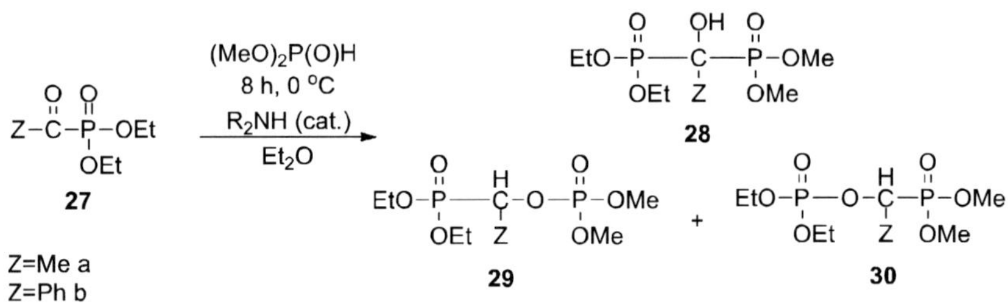

Scheme 10. The reaction of diethyl α-oxophosphonates with dimethyl phosphite [7].

For Z = Me using 5% of DEA as a catalyst, the corresponding adduct (**28a**) was formed in 94%, and the two rearranged products (**29a** and **30a**) in 4% and 2%, respectively (Table 6, Entry 1). With the addition of 40% DEA, almost complete rearrangement accompanied the reaction, the ratio of the three products (**28a**:**29a**:**30a**) was 3:76:21 (Table 6, Entry 2). The reaction was also carried out in the presence of 5% DBA. In this case, only the adduct (**28a**) was formed (Table 6, Entry 3) [7].

The reaction of diethyl benzoylphosphonate and dimethyl phosphite was carried out in the presence of 5% DEA. In this case, only the two rearranged products (**29b**:**30b**) were formed in a ratio of 81:19 (Table 6, Entry 4) [7].

During the reaction of diethyl benzoylphosphonate (**27b**) and dimethyl phosphite, the tendency for rearrangement is much larger due to the presence of the phenyl substituent. Even in the presence of a small amount of catalyst, the rearranged products (**29b** and **30b**) were formed. In this case, the adduct (**28b**) is a real intermediate [7].

Oxophosphonates (**27**) were also reacted with diphenylphosphine oxide (Scheme 11).

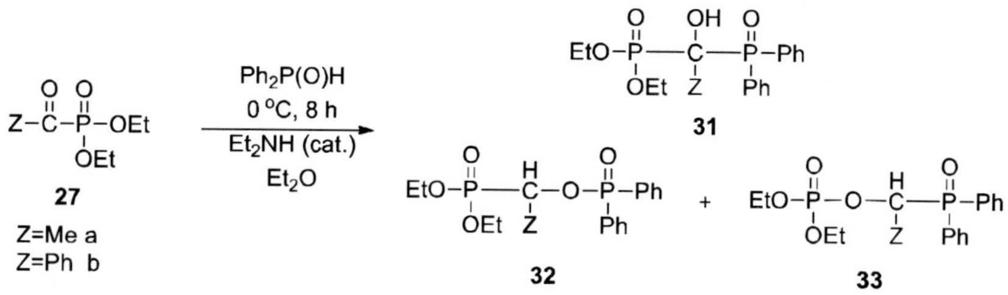

Scheme 11. The reaction of diethyl α-oxophosphonates with diphenylphosphine oxide [7].

Table 7. The reaction conditions of diethyl α-oxophosphonates and diphenylphosphine oxide [7]

Exp.	Z	Catalyst (%)	Product composition (%) [a]			Yield (%)
			31	32	33	
1.	Me	20	100	-	-	74 (31a)
2.	Ph	20	-	60	40	81 (32b+33b)

[a] Based on the relative ^{31}P NMR intensities.

In the case of the reaction of diethyl acetylphosphonate (**27a**) with diphenylphosphine oxide in the presence of 20% diethylamine, the Pudovik adduct (**31a**) was formed (Table 7, Entry 1). On the other hand, by performing the reaction with diethyl benzoylphosphonate (**27b**) the 60:40 mixture of the two possible rearranged products (**32b** and **33b**) was obtained (Table 7, Entry 2) [7].

The formation of rearranged derivatives at such a rate can be explained by the electrophilicity of the $Ph_2P(O)$ and $(EtO)_2P(O)$ molecular parts. Due to the steric hindrance ($Ph_2P(O)$ part), the two isomers (**32b** and **33b**) are formed in a more balanced manner (40-60%), in contrast to the previously described case (**29b** and **30b**) (20-80%). The significant rearrangement can be explained by the energy of 385 kJ/mol^{-1} released during the formation of the P–O bond. If the substituent attached to the central carbon atom is a phenyl group, the adduct is less stable than in the case of a methyl substituent [7, 38].

The Pudovik reaction between α-oxophosphonates and dialkyl phosphites or diphenylphosphine oxide ($Y_2P(O)H$; Y = OMe, OEt, Ph) results the corresponding adducts and/or their rearranged derivatives. The reactions are mainly influenced by the Z substituent and the amount of secondary amine.

The quality of the amine, the temperature, the solvent and the Y substituent have a smaller influence. Time proved to have a special effect [7].

Preparation of Bisphosphonate Derivates by Other Synthetic Possibilities

Migianu and her group mentioned a selective one-pot synthesis of partial dimethyl esters of hydroxy-methylenebisphosphonic acid (Scheme 12). The reaction of acid chloride (**17**) with 2 equivalents of methyl bis(trimethylsilyl) phosphite at room temperature afforded the product (**34**). The synthesis was based on the Arbuzov reaction [39].

McKenna et al., reported the preparation of α-hydroxy-alkyl/aryl-methylenebisphosphonates (**36**) by Grignard reaction from tetraalkyl carbonyl-bisphosphonates (**35**) (Scheme 13) [40].

Bisphosphonate derivatives, which do not contain a hydroxy group on their α-carbon atom, can be prepared by the reaction of diiodomethane (**37**) and trialkyl phosphite. First, a double Arbuzov reaction takes place, and then the tetraalkyl bisphosphonate (**38**) is alkylated in the presence of an alkali metal hydride base with the appropriate alkylating agent in dimethyl-formamide solvent. Thus, α-substituted methylenebisphosphonic acid ester (**39**) will be the resulting product (Scheme 14) [41, 42].

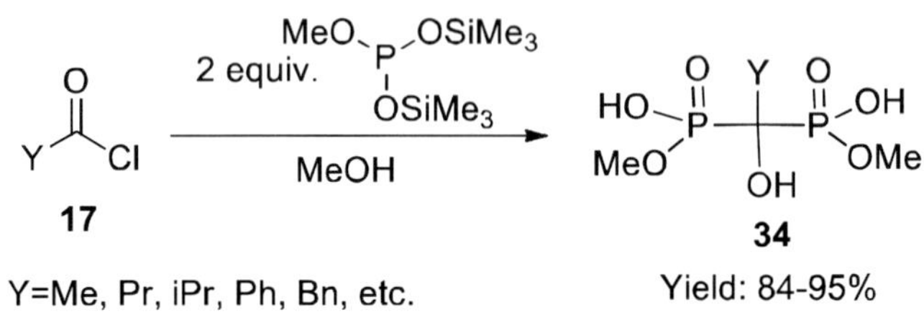

Scheme 12. One-pot synthesis of partial dimethyl esters of hydroxy-methylenebisphosphonic acid [39].

Similarly, when lithiated dialkylphosphonate (**41**) is reacted with dialkyl 180hlorophosphite, α-substituted methylenebisphosphonic acid ester (**39**) is also obtained as a product (Scheme 15) [43].

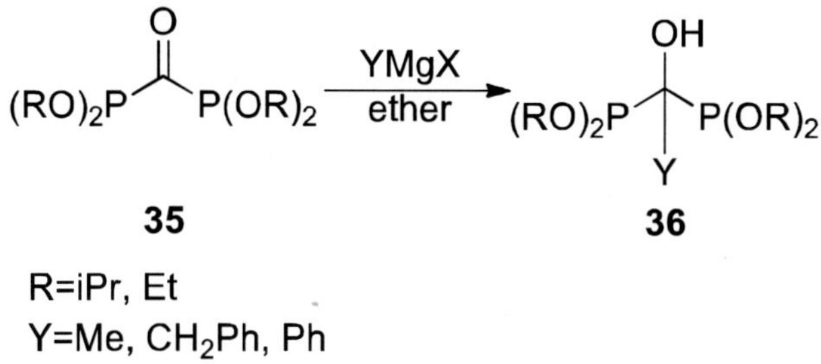

Scheme 13. Preparation of α-hydroxy-alkyl/aryl-methylenebisphosphonates by Grignard reaction [40].

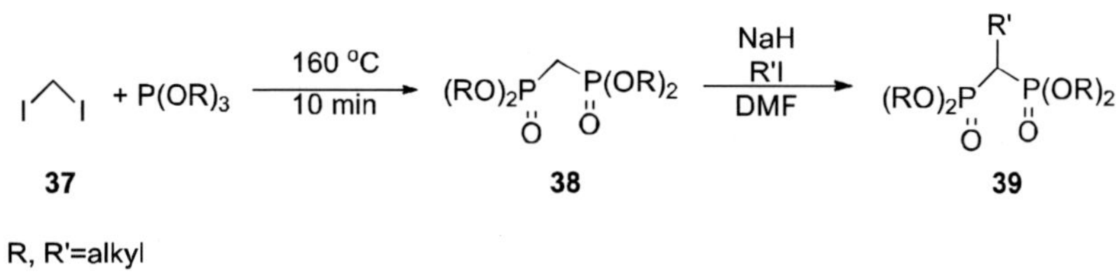

Scheme 14. Preparation of bisphosphonate derivatives using diiodomethane and trialkylphosphite [41, 42].

Ortial et al., prepared bisphosphorus compounds by the reaction of the anion of a phosphonate diester with a P(III)-chlorophosphine or its P(V)-borane complex (Scheme 16) The received products (**42**) can be reacted with alkylating agent or carbonyl compounds after deprotonation [44].

Antczak and Montchamp first described a facile synthesis of P–C–P compounds (**48** and **49**) based on the reaction of phosphorus carbenoids with organoboranes, followed by a reaction with phosphorus electrophiles (Scheme 17) [45].

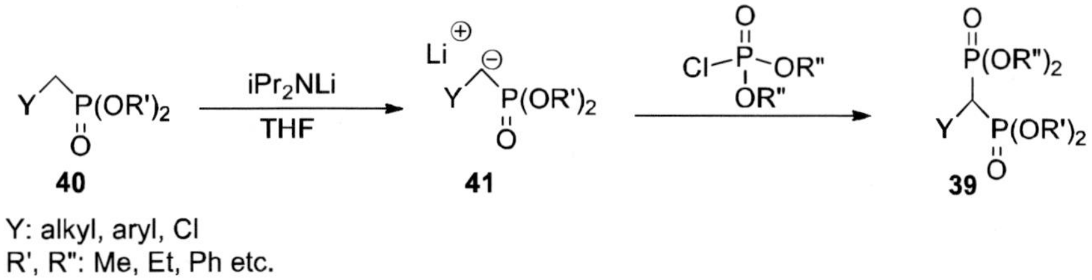

Scheme 15. The reaction of lithiated dialkylphosphonates and dialkyl chlorophosphates to form α-substituted methylenebisphosphonic acid esters [43].

Byers and co-workers synthesized geminal bisphosphonates (**51**) to various monosubstituted alkenes by photochemical radical addition of tetraethyl phenylseleno-methylenebisphosphonate (**50**) (Scheme 18). The phenylselenide (**50**) was prepared by deprotonation of tetraethyl methylenebisphosphonate (**38**) with NaH followed by the addition of PhSeCl [46].

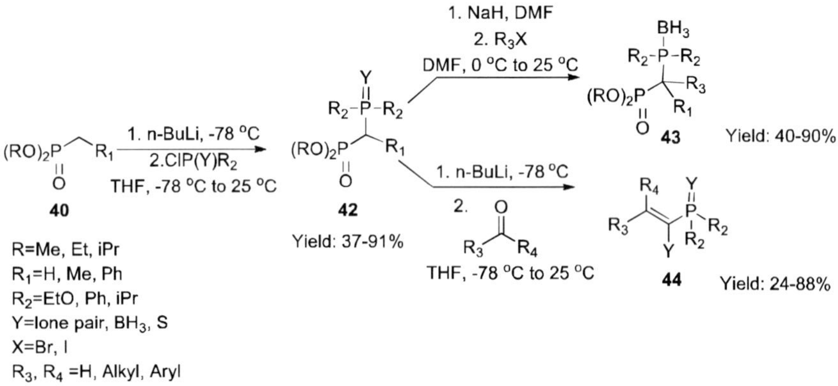

Scheme 16. Reaction of a phosphonate diester anion with a P(III)-chlorophosphine or its P(V)-borane complex [44].

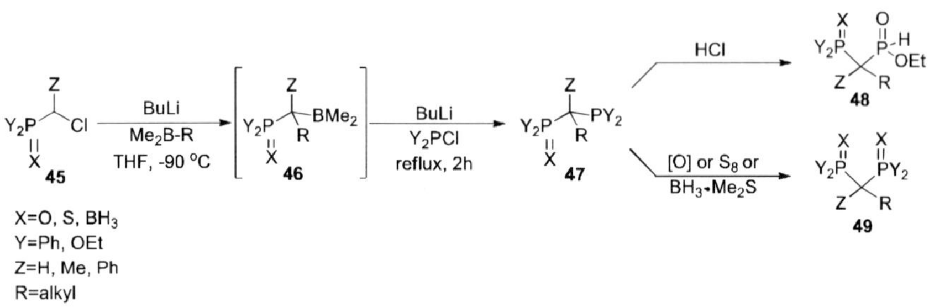

Scheme 17. Synthesis of P–C–P compounds based on the reaction of phosphorus carbenoids with organoboranes [45].

The research group of Haelters reported the synthesis of a series of new bisphosphonate compounds (**53**), in which there is an alkoxy group on the central carbon atom, which probably increases their lipophilicity and thus their bioavailability. Alkylation of these compounds with allyl bromide was also studied (Scheme 19) [47].

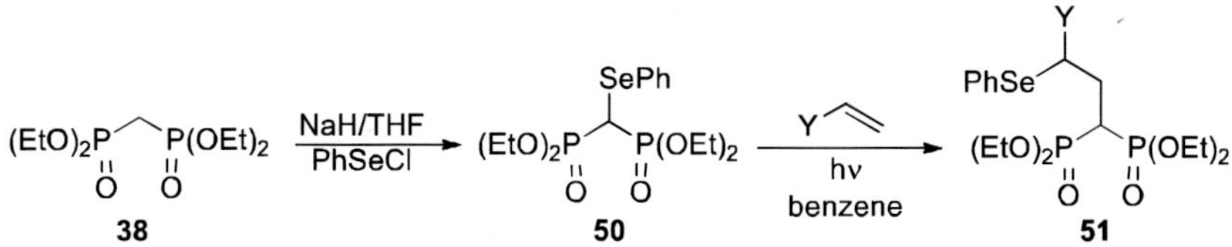

Y=Hex, OBu, OAc, etc.

Scheme 18. Preparation of geminal bisphosphonates by photochemical radical addition [46].

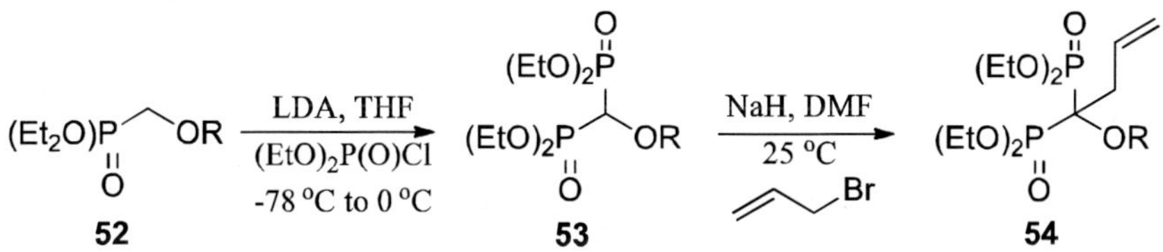

R=Me, Bu, Hex, CH_2Ph, etc.

Scheme 19. Synthesis of functionalized alkoxyalkylidene gem-bisphosphonates [47].

Gajda and his research group described a one-pot synthesis of substituted tetraethyl N-Boc 2-aminoethylidene-1,1-bisphosphonates (**58**). The method involves the base-induced addition of the sodium salt of tetraethyl methylenebisphosphonate (**56**) to N-Boc imines (**57**). This compound (**57**) was prepared *in situ* from N-Boc-α-amidoalkyl-p-tolylsulfones (**55**) with sodium hydride (Scheme 20) [48].

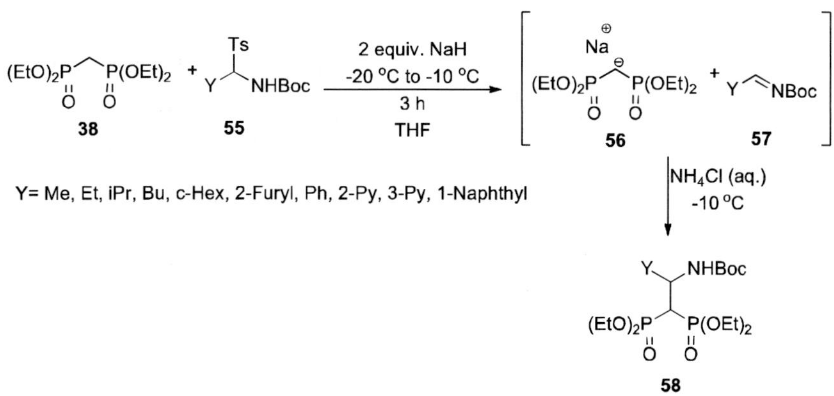

Y= Me, Et, iPr, Bu, c-Hex, 2-Furyl, Ph, 2-Py, 3-Py, 1-Naphthyl

Scheme 20. Synthesis of substituted tetraethyl N-Boc 2-aminoethylidene-1,1-bisphosphonates [48].

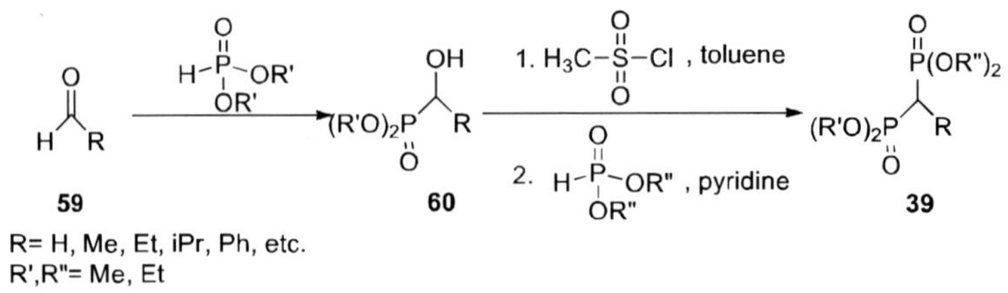

Scheme 21. Synthesis of tetraalkyl bisphosphonate esters from α-hydroxyphosphonate [4].

According to an alternative synthetic method, α-hydroxyphosphonate (**60**) is formed from aldehydes (**59**) in the Pudovik reaction with dialkyl phosphite. By reacting the resulting product (**60**) with mesyl chloride (to create a better leaving group), and then adding another dialkyl phosphite, the tetraalkyl bisphosphonate esters (**39**) are obtained (Scheme 21) [4].

Heuzeé's co-workers reported that adding an excess of the sodium salt of diisopropyl phosphite (**61**) to carbon disulfide at room temperature in an aprotic solvent results complete desulfurization of the latter, with the formation of sodium salt of methylenebisphosphonate (**62**) and sodium thiophosphate (**63**). After hydrolysis diisopropyl methylenebisphosphonate (**64**) and thiophosphorous acid (**65**) are produced (Scheme 22) [49].

Pieper et al., described the synthesis of tetraethyl hydroxyphenyl-methylenebisphosphonate (**71**) by hydroxy-de-diazoniation of the corresponding diazonium salt of tetraethyl aminophenyl-methylene-bisphosphonate (**70**) (Scheme 23) [50].

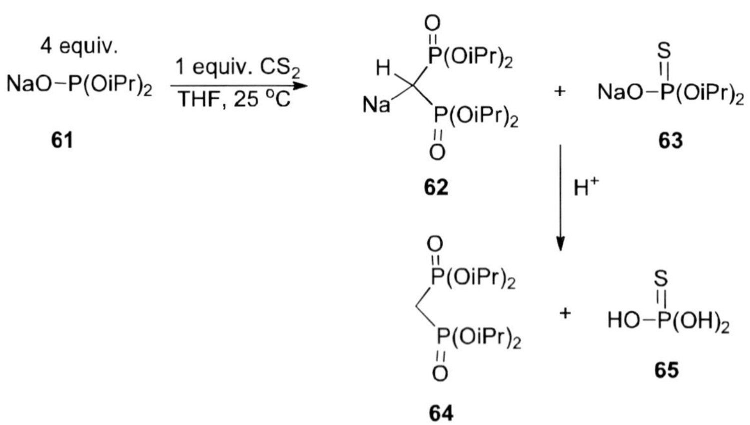

Scheme 22. The reaction of the sodium salt of diisopropyl phosphite and CS_2 [49].

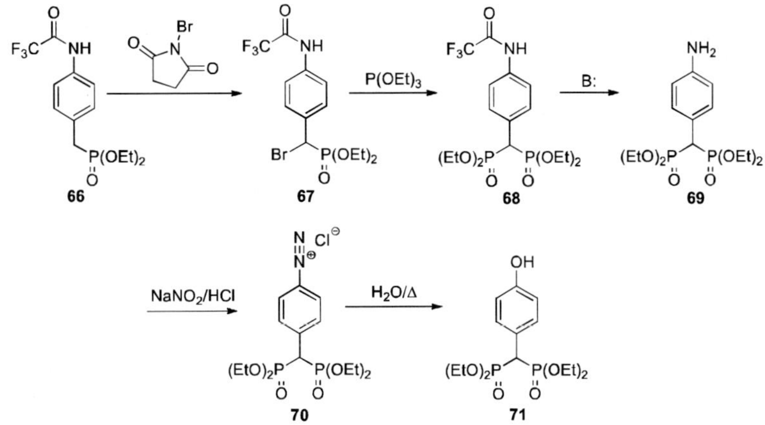

Scheme 23. Synthesis of tetraethyl hydroxyphenyl-methylenebisphosphonate by hydroxy-de-diazoniation [50].

Conclusion

The most commonly used solution for the synthesis of α-hydroxy-methylenebisphosphonates in the industry is the so-called "direct" synthesis, in which case the corresponding amino acids, heteroaryl acetic acids or derivatives are reacted with various P-reagents. The disadvantages of this method are the extremely heterogenous reaction of the mixture, viscousity, difficulties in mixing and the released large amount of hydrochloric acid gas by the hydrolysis. These problems are manageable in laboratory environment, but it is difficult to perform these transformations in industrial environment and to keep these reactions in safe condition. Another possibility for the preparation of these compounds is the "indirect" synthesis. In this case, in certain conditions (for instance catalyst, temperature) the desired product and their rearranged deriatives are formed, too. There are many different possibilities for the synthesis of (hydroxy-)methylenebisphosphonates. Some of these have been illustrated in this chapter but not exhaustively.

Acknowledgments

The project was supported by Gedeon Richter Talentum Foundation established by Gedeon Richter Plc. (1103 Budapest, Gyömrői út 19-21.) with the support of the Gedeon Richter Excellence PhD Scholarship.

Zsuzsanna Szalai is offering her thanks to professor György Keglevich DSc. and all co-authors appearing in the cited references.

Disclaimer

The author declares no conflict of interest.

References

[1] Mucha, A., Kafarski, P., Berlicki, Ł. Remarkable Potential of the α-Aminophosphonate/Phosphinate Structural Motif in Medicinal Chemistry. *J. Med. Chem.* 2011, 54, 5955–5980, doi: https://doi.org/10.1021/jm200587f.

[2] Hudson, H. R., Wardle, N. J., Bligh, S. W. A., Greiner, I., Grün, A., Keglevich, G. N-Heterocyclic Dronic Acids: Applications and Synthesis. *Mini-Reviews Med. Chem.* 2012, 12, 313–325, doi: https://doi.org/10.2174/138955712799829285.

[3] Nagy, D. I., Grün, A., Greiner, I., Keglevich, G. The Role of Phosphorus Trichloride and Phosphorous Acid in the Formation of Alpha-Hydroxymethylenebisphosphonic Acids from the Corresponding Carboxylic Acids – A Mechanistic Overview. *Curr. Org. Chem.* 2017, 21, 1567–1578, doi: https://doi.org/10.2174/1385272821666170417122441.

[4] Li, C., Yuan, C. Studies on Organophosphorus Compounds 68. A New and Facile Synthetic Approach to Alkylidenebisphosphonates. *Tetrahedron Lett.* 1993, 34, 1515–1516, doi: https://doi.org/10.1016/S0040-4039(00)60333-7.

[5] Nicholson, D. A., Vaughn, H. A General Method of Preparation of TetramethylAlkyl-1-Hydroxy-1,1-Diphosphonates. *J. Org. Chem.* 1971, 36, 3843–3845, doi: https://doi.org/10.1021/jo00823a621.

[6] Grün, A., Molnar, I. G., Bertok, B., Greiner, I., Keglevich, G. Synthesis of α-Hydroxy-Methylenebisphosphonates by the Microwave-Assisted Reaction of α-Oxophosphonates and Dialkyl Phosphites under Solventless Conditions. *Heteroatom Chem.* 2009, 20, 350–354, doi: https://doi.org/10.1002/hc.20558.

[7] Szalai, Z., Keglevich, G. Tetraalkyl Hydroxymethylene-Bisphosphonate and Dialkyl 1-Diphenylphosphinoyl-1-Hydroxy-Ethylphosphonate Derivatives by the Pudovik Reaction and Their Rearranged Products. *Molecules* 2021, 26, 7575, doi: https://doi.org/10.3390/molecules26247575.

[8] Gyíres, K., Fürst, Z. *A Farmakológia Alapjai (Foundations of Pharmacology)*; Medicina Publishing Co. Ltd., 2011; ISBN 978 963 226 324 3.

[9] Owen, R., Reilly, G. C. *In Vitro* Models of Bone Remodelling and Associated Disorders. *Front. Bioeng. Biotechnol.* 2018, 6, 1–22, doi: https://doi.org/10.3389/fbioe.2018.00134.

[10] Trzeciakiewicz, A., Habauzit, V., Horcajada, M. N. When Nutrition Interacts with Osteoblast Function: Molecular Mechanisms of Polyphenols. *Nutr. Res. Rev.* 2009, 22, 68–81, doi: https://doi.org/10.1017/S095442240926402X.

[11] Widler, L., Jaeggi, K. A., Glatt, M., Müller, K., Bachmann, R., Bisping, M., Born, A. R., Cortesi, R., Guiglia, G., Jeker, H., Rémy Klein, Ueli Ramseier, Johann Schmid, Gerard Schreiber, Yves Seltenmeyer, Jonathan R. Green. Highly Potent Geminal Bisphosphonates. From Pamidronate Disodium (Aredia) to Zoledronic Acid (Zometa). *J. Med. Chem.* 2002, 45, 3721–3738, doi: https://doi.org/10.1021/jm020819i.

[12] Russell, R., Graham, G. Bisphosphonates: The First 40 Years. *Bone* 2011, 49, 2–19, doi: https://doi.org/10.1016/j.bone.2011.04.022.

[13] Catterall, J. B., Cawston, T. E. Drugs in Development: Bisphosphonates and Metalloproteinase Inhibitors. *Arthritis Res. Ther.* 2002, 5, 12–24, doi: https://doi.org/10.1186/ar604.

[14] Fischer, J., Ganellin, C. R. *Analogue-based Drug Discovery*; WILEY-VCH Verlag GmbH & Co. KGaA, 2006; ISBN 3527608001.

[15] Keglevich, G., Grün, A.; Aradi, K., Garadnay, S., Greiner, I. Optimized Synthesis of N-Heterocyclic Dronic Acids; Closing a Black-Box Era. *Tetrahedron Lett.* 2011, 52, 2744–2746, doi: https://doi.org/10.1016/j.tetlet.2011.03.093.

[16] Nagy, D. I. Biszfoszfonátok Előállításának Vizsgálata. *Magyar Kémiai Folyóirat* 2018 (Investigation of the Production of Bisphosphonates. Hungarian *Chemistry Journal* 2018), 124, 6–12, doi: https://doi.org/10.24100/mkf.2018.01.6.

[17] Grün, A., Szalai, Z., Keglevich, G. "Greener" Synthesis of Zoledronic Acid from Imidazol-1-Yl-Acetic Acid and P-Reagents Using Diethyl Carbonate as the Solvent Component. *Lett. Org. Chem.* 2021, 18, 8–12, doi: https://doi.org/10.2174/1570178617999200730203738.

[18] Kieczykowski, G. R., Jobson, R. B., Melillo, D. G., Reinhold, D. F., Grenda, V. J., Shinkai, I. Preparation of (4-Amino-l-Hydroxybutylidene)Bisphosphonic Acid Sodium Salt, MK-217 (Alendronate Sodium). An Improved Procedure for the Preparation of 1-Hydroxy-1,1-Bisphosphonic Acids. *J. Org. Chem.* 1995, 60, 8310–8312, doi: https://doi.org/10.1021/jo00130a036.

[19] Nagy, D. I., Grün, A., Lévay, K., Garadnay, S., Greiner, I., Keglevich, G. Efficient Syntheses of Zoledronic Acid as an Active Ingredient of a Drug against Osteoporosis. *Synth. Commun.* 2018, 48, 663–671, doi: https://doi.org/10.1080/00397911.2017.1410894.

[20] Mustafa, D. A., Kashemirov, B. A., McKenna, C. E. Microwave-Assisted Synthesis of Nitrogen-Containing 1- Hydroxymethylenebisphosphonate Drugs. *Tetrahedron Lett.* 2011, 52, 2285–2287, doi: https://doi.org/10.1016/j.tetlet.2011.02.058.

[21] Pandey, S. C., Haider, H., Saxena, S., Singh, M. K., Thaper, R. K., Dubey, S. K. *Process for Producing Bisphosphonic Acids and Forms Thereof* 2009.

[22] Rao, D. V. N. S., Dandala, R., Narayanan, G. K. A. S. S., Lenin, R., Sivakumaran, M., Naidu, A. Novel Procedure for the Synthesis of 1-Hydroxy-1,1-Bisphosphonic Acids Using Phenols as Medium. *Synth. Commun.* 2007, 37, 4359–4365, doi: https://doi.org/10.1080/00397910701578545.

[23] Kovács, R., Grün, A., Németh, O., Garadnay, S., Greiner, I., Keglevich, G. The Synthesis of Pamidronic Derivatives in Different Solvents: An Optimization and a Mechanistic Study. *Heteroatom Chem.* 2014, 25, 186–193, doi: https://doi.org/10.1002/hc.21155.

[24] Grün, A., Nagy, D. I., Németh, O., Mucsi, Z., Garadnay, S., Greiner, I., Keglevich, G. The Synthesis of 3-Phenylpropidronate Applying Phosphorus Trichloride and Phosphorous Acid in Methanesulfonic Acid. *Curr. Org. Chem.* 2016, 20, 1745–1752, doi: https://doi.org/10.2174/1385272820666160427115559.

[25] Nguyen, L. M., Niesor, E., Bentzen, C. L. Gem-Diphosphonate and Gem-Phosphonate-Phosphate Compounds with Specific High Density Lipoprotein Inducing Activity. *J. Med. Chem.* 1987, 30, 1426–1433, doi: https://doi.org/10.1021/jm00391a027.

[26] Boppudi, H. B., Gandavaram, S. P., Chamarthi, N. R., Mandava, V. B. R. Synthesis of Phosphonates via Michaelis-Arbuzov Reaction. *Curr. Org. Synth.* 2017, 14, 883–903, doi: https://doi.org/10.2174/1570179414666161230144455.

[27] Keglevich, G., Bálint, E. The Kabachnik-Fields Reaction: Mechanism and Synthetic Use. *Molecules* 2012, 17, 12821–12835, doi: https://doi.org/10.3390/molecules171112821.

[28] Ali, T. E., Abdel-Kariem, S. M. Methods for the Synthesis of α-Heterocyclic/Heteroarylα-Aminophosphonic Acids and Their Esters. *Arkivoc* 2015, 246–287, doi: https://doi.org/10.3998/ark.5550190.p009.112.

[29] Engel, R. Phosphorus Addition at Sp2 Carbon. *Organic Reactions* 1988, 175–248, doi: https://doi.org/10.1002/0471264180.or036.02

[30] Fields, E. K. The Synthesis of Esters of Substituted Amino Phosphonic Acids. *J. Am. Chem. Soc.,* 1952, 74, 1528–1531, doi: https://doi.org/10.1021/ja01126a054.

[31] McConnell, R. L., Coover, W. Preparation of 1-Hydroxyalkylidenediphosphonates. *J. Am. Chem. Soc.,* 1956, 78, 4450–4452, doi: https://doi.org/10.1021/ja01598a069.

[32] Fitch, S. J., Moedritzer, K. N. M. R. Study of the P-C(OH)-P to P-C-O-P Rearrangement: Tetraethyl 1-Hydroxy Alky Lidenediphosphonates. *J. Am. Chem. Soc.,* 1962, 84, 1876–1879, doi: https://doi.org/10.1021/ja00869a022.

[33] Turhanen, P. A., Ahlgren, M. J., Järvinen, T., Vepsäläinen, J. J. Bisphosphonate Prodrugs. Selective Synthesis of (1-Hydroxyethylidene)-1,1-Bisphosphonate Partial Esters. *Synthesis (Stuttg).* 2000, 4, 633–637, doi: https://doi.org/10.1055/s-2001-12353.

[34] Turhanen, P. A., Ahlgren, M. J., Järvinen, T., Vepsäläinen, J. J. Bisphosphonate Prodrugs. Synthesis and Identification of (1-Hydroxyethylidene)-1,1-Bisphosphonic Acid Tetraesters by Mass Spectrometry, NMR Spectroscopy and x-Ray Crystallography. *Phosphorus, Sulfur Silicon Relat. Elem.* 2001, 170, 115–133, doi: https://doi.org/10.1080/10426500108040589.

[35] Prechelmacher, S., Mereiter, K., Hammerschmidt, F. The α-Hydroxyphosphonate-Phosphate Rearrangement of a Noncyclic Substrate-Some New Observations. *Org. Biomol. Chem.* 2018, 16, 3672–3680, doi: https://doi.org/10.1039/c8ob00419f.

[36] Pallitsch, K., Roller, A., Hammerschmidt, F. The Stereochemical Course of the α-Hydroxyphosphonate-Phosphate Rearrangement. *Chem. - A Eur. J.* 2015, 21, 10200–10206, doi: https://doi.org/10.1002/chem.201406661.

[37] Grün, A., Molnar, I. G., Greiner, I., György, K. Phenyl-, Benzyl-, and Unsymmetrical Hydroxy-Methylenebisphosphonates as Dronic Acid Ester Analogues from α-Oxophosphonates by Microwave-Assisted Syntheses. *Heteroatom Chem.* 2011, 22, 640–680, doi: https://doi.org/10.1002/hc.20727.

[38] Hartley, S. B., Holmes, W. S., Jacques, J. K., Mole, M. F., Mccoubrey, J. C. Thermochemical Properties of Phosphorus Compounds. *Q. Rev. Chem. Soc.* 1963, 17, 204–223, doi: https://doi.org/10.1039/qr9631700204.

[39] Migianu, E., Mallard, I., Bouchemal, N., Lecouvey, M. One-Pot Synthesis of 1-Hydroxymethylene-1,1-Bisphosphonate Partial Esters. *Tetrahedron Lett.* 2004, 45, 4511–4513, doi: https://doi.org/10.1016/j.tetlet.2004.04.047.

[40] McKenna, C. E., Kashemirov, B. A., Li, Z. M. Synthetic Approaches to Biologically Active Bisphosphonates and Phosphonocarboxylates. *Phosphorus, Sulfur Silicon Relat. Elem.* 1999, 144–146, 313–316, doi: https://doi.org/10.1080/10426509908 546244.

[41] Kosolapoff, G. M. The Chemistry of Aliphatic Phosphonic Acids. I. Alkylation of Methanediphosphonic Acid. *J. Am. Chem. Soc.* 1953, 75, 1500–1501, doi: https:// doi.org/10.1021/ja01102a524.

[42] Magnin, D. R., Biller, S. A., Dickson, J. K., Logan, J. V., Lawrence, R. M., Chen, Y., Sulsky, R. B., Ciosek, C. P., Harrity, T. W., Jolibois, K. G., Lori K. Kunselman, Lois C. Rich, and Dorothy A. Slusarchyk. 1,1- Bisphosphonate Squalene Synthase Inhibitors: Interplay Between the Isoprenoid Subunit and the Diphosphate Surrogate. *J. Med. Chem.* 1995, 38, 2596–2605, doi: https://doi.org/10.1021/jm000 14a012.

[43] Teulade, M. P., Savignac, P., Aboujaoude, E. E., Liétge, S., Collignon, N. Alkylidènediphosphonates et Vinylphosphonates: Une Démarche Synthétiques Sélective Par Voie Carbanionique. *J. Organomet. Chem.* 1986, 304, 283–300, doi: https://doi.org/10.1016/0022-328X(86)80147-4.

[44] Ortial, S., Thompson, D. A., Montchamp, J. L. Mixed 1,1-Bisphosphorus Compounds: Synthesis, Alkylation, and Horner-Wadsworth-Emmons Olefination Reactions. *J. Org. Chem.* 2010, 75, 8166–8179, doi: https://doi.org/10.1021/jo 101814w.

[45] Antczak, M. I., Montchamp, J. L. Synthesis of 1,1-Bis-Phosphorus Compounds from Organoboranes. *Tetrahedron Lett.* 2008, 49, 5909–5913, doi: https://doi.org/ 10.1016/j.tetlet.2008.07.144.

[46] Byers, J. H., Thissell, J. G., Thomas, M. A. The Synthesis of Geminal Diphosphonates via Phenyl Selenide Transfer Radical Addition. *Tetrahedron Lett.* 1995, 36, 6403–6406, doi: https://doi.org/10.1016/0040-4039(95)01324-B.

[47] Haelters, J. P., Couthon-Gourvès, H., Le Goff, A., Simon, G., Corbel, B., Jaffrès, P. A. Synthesis of Functionalized Alkoxyalkylidene Gem-Bisphosphonates. *Tetrahedron* 2008, 64, 6537–6543, doi: https://doi.org/10.1016/j.tet.2008.04.052.

[48] Gajda, A., Gajda, T. A New Access to Substituted Tetraethyl N-Boc 2-Aminoethylidene-1,1-Bisphosphonates and Phosphonyl-Substituted Aza-Morita-Baylis-Hillman-Type Adducts. *Tetrahedron* 2008, 64, 1233–1241, doi: https://doi. org/10.1016/j.tet.2007.11.072.

[49] Heuzé, B., Lemarié, M., Vazeux, M., Gulea, M., Masson, S., Sene, A., Jaffrès, P. A., Alberti, A., MacCiantelli, D. Synthesis of Methylene Bisphosphonates from Carbon Disulfide and Phosphites via Desulfurization: A Mechanistic Study. *Phosphorus, Sulfur Silicon Relat. Elem.* 2009, 184, 820–829, doi: https://doi.org/10.1080/10426500802715528.

[50] Pieper, T., Keppler, B. K. Preparation of Tetraethyl-4-Hydroxyphenylmethylene-1,1-Bisphosphonate by Hydroxy-de-Diazoniation of the Corresponding Diazonium Salt of Tetraethyl-4-Aminophenylmethylene-1,1-Bisphosphonate. *Phosphorus, Sulfur Silicon Relat. Elem.* 2000, 165, 77–82, doi: https://doi.org/10.1080/10426500008076326.

Chapter 7

The Exploration of Greener Technologies in the Synthesis of Anti-COVID-19 Drugs

Trinath Biswal[1]
D. S. Ramakrishna[1]
and K. Sridhar Reddy[2]

[1]Department of Chemistry, Veer Surendra Sai University of Technology, Burla, Sambalpur, Odisha, India
[2]Department of Chemistry, Palamuru University Mahabubnagar, Telangana, India

Abstract

To maintain sustainable development of our globe, organic materials should be synthesized by using the twelve principles of green chemistry. Due to the generation of different variants of COVID 19 (Corona virus) from time to time, the need for anti-COVID-19 drugs is vastly increasing. There is a great challenge in this situation to meet this challenge. We must develop a methodology of preparing anti-COVID-19 drugs based on eco-friendly processes that should satisfy pharmaceutical industries. Today, most of the synthesis process proceeds with several reactions associated with numerous toxic chemicals in the form of substrates, reagents, catalysts, additives, and solvents. Various challenges were faced during the storage, quenching, and transportation of toxic chemical substances. Since most of the chemicals used are not biodegradable, it is a serious threat to our ecosystem and biological community of the globe. Hence, to solve this global crisis, numerous green technologies, such as less hazardous chemical synthesis process must be adopted, leading to the use of bio-resources and water as solvents, one-pot synthesis, etc. The method of one-pot synthesis is a chemical reaction that is carried out in a single reactor, and no purification of the intermediate products is

In: The Future of Green Synthesis
Editor: Jacques Kent
ISBN: 979-8-88697-816-2

required. For organic intermediate purification, huge quantities of toxic organic solvents, such as chlorinated solvents including CH_2Cl_2, $CHCl_3$, and CCl_4 and highly flammable solvents such as hexane and benzene, are required. Hence, one-pot synthesis is not only a green pathway of synthesis but is also cost-effective. Therefore, to minimize the pollution level in the environment, various green technology pathways should be developed for the synthesis of drugs and organic compounds. The present chapter summarizes remarkable green chemistry initiatives in anti-COVID-19 drugs based on organic molecules. The drugs used for COVID-19 treatment are highly effective for treatment of viral diseases that primarily affect the respiratory system, where an epithelial cell of the respiratory system is damaged, resulting in mucosal membrane inflammation. Inflammatory mucosal membranes lead to alveolar damage and eventually pneumonia. Covid-19 is caused by a positive-sense single-stranded RNA virus, SARS-CoV-2, usually known as the novel corona virus. Even though many drugs have been screened and tested in clinical trials, only a few of them, such as hydroxychloroquine, favipiravir, remdesivir, molnupiravir, etc., have attracted scientists and shown remarkable activity against SARS-CoV-2.

Introduction

Based on the report of the WHO, up to February 9, 2023, it was reported that 755,166,409 confirmed cases of COVID-19, including 6,831,681 deaths were observed worldwide. AS per the data of WHO up to December 12, 2022, a total of 13,008,560,983 vaccine doses have been administered worldwide [1]. In December 2019, the novel severe acute respiratory syndrome coronavirus 2 (SARS-CoV-2), the causative agent of coronavirus disease 2019 (COVID-19), was first identified in the province of Wuhan, China. Since then, there have been over 400 million confirmed cases and 5.8 million deaths from COVID-19 reported worldwide. There is an urgent need for therapies against SARS-CoV-2 led researchers to use drug repurposing approaches. This strategy allows for the reduction of risks, time, and costs associated with drug development. In many cases, a repurposed drug can enter directly into preclinical testing and clinical trials, thus accelerating the whole drug discovery process. In this work, we will give a general overview of the main developments in COVID-19 treatment, focusing on the contribution of the drug repurposing paradigm to finding effective drugs against this disease. Although powerful vaccines were established in this regard, there was still a need for pharmacological approaches against COVID-19. These alternative approaches were found to be inevitable, and it is necessary to recognize and expand novel medicine for COVID-19 management, and thereby it will be easier to reduce the following:

- The troubles of diseases
- The number of hospital admissions
- Mortality.

Conventional treatments rely on identifying a novel selective antiviral molecule along with the advancement of a novel therapeutic, this will be (i) a time-consuming procedure that generally takes decades, and (ii) an expensive task. As an alternative to this, drug repurposing was evolved as a constructive and highly reliable method to recognize the existing approved medicines to cure different kinds of viruses, in connection with COVID-19, and to overcome the emergency created by this dangerous virus. The major benefit of drug repurposing relies on the accessibility of data about the (i) pharmacokinetics, (ii) pharmacodynamics, and (iii) toxicity, of a given medicine prior to its use [2, 3]. By means of analogous methodologies, better anti-SARS-CoV-2 drugs could extensively reduce the time required to discover a successful treatment procedure to cure COVID-19, in terms of (i) dropping disease troubles, and (ii) reducing confirmed cases and deaths. Starting from the evolution of the pandemic, a variety of effective anti-SARS-CoV-2 medicines were examined at various stages of clinical trials, for example, (i) remdesivir (at first established by Gilead), which was used to treat Hepatitis C, (ii) chloroquine and hydroxychloroquine, which were found to be useful to treat malaria, (iii) molnupiravir (Merck), which was used to treat influenza viruses and encephali [4].

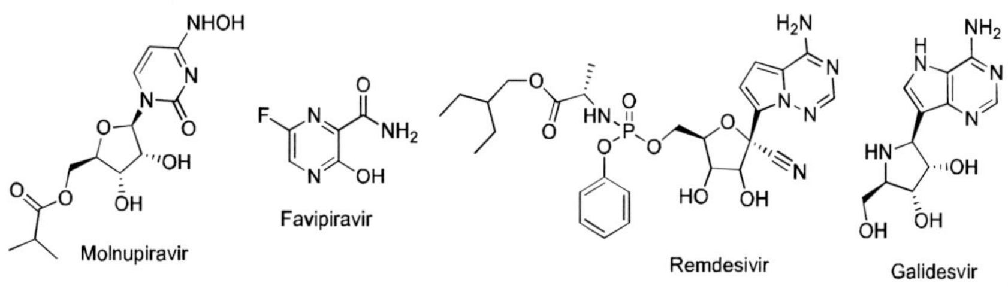

Figure 1. Structures of Remdesivir and other related anti viral drugs.

Table 1. Most studied repurposed medicines to treat COVID-19 [6]

S. No	Repurposed medicine	Original Approved Therapeutic Use
1	Remdesivir	Ebola virus
2	Hydroxychloroquine	Malaria
3	Molnupiravir	Influenza viruses and encephalitic alpha viruses
4	Favipiravir	Influenza virus
5	Galidesvir	Hepatitis C

The common factor in the action of these simple molecules or antiviral drugs is their probable mechanism of action, which has been found to be similar in curing COVID-19, i.e., all these molecules are able to inhibit virus RNA synthesis [5]. Table 1 depicts various potential COVID-19 treatment medications.

Molnufiravir

Merck developed molnupiravir, marketed as "LagevrioTM," and it was discovered to be the first oral antiviral that directly treats nasopharyngeal SARS-CoV-2 titers. In addition to this, in cases of mild COVID-19 stages, a noteworthy advantage is observed in decreasing hospital admissions or mortality rates (up to 30%) [7, 8]. It belongs to the subcategory of prodrugs; when molupiravir interacts with various host kinases, it will be converted from a hydroxyl functional group containing cytidine into its corresponding 5'-triphosphate, that will have the required biological activities. Based on the COVID-19 stages, both the EMA and the FDA were granted and authorized to use this drug [9, 10]. Molnupiravir can be obtained from different naturally occurring starting materials such as uridine, cytidine, ribose, and uracil (Figure 2).

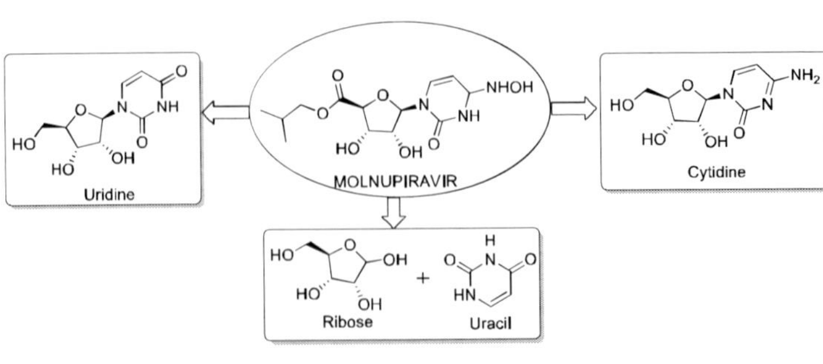

Figure 2. Retro-synthetic approaches for Molnupiravir.

The interesting story in achieving a sustainable synthetic strategy aimed at Molnupiravir (MK-4482 or EIDD-2801) starts with the initial route reported, which requires 10 steps and a 10% overall yield reported by employing ribose and uracil as starting materials. Huge amounts of chemicals, such as organic solvents, protecting group reagents, were required to complete the synthesis in a bulk manner (Scheme 1) [11, 12].

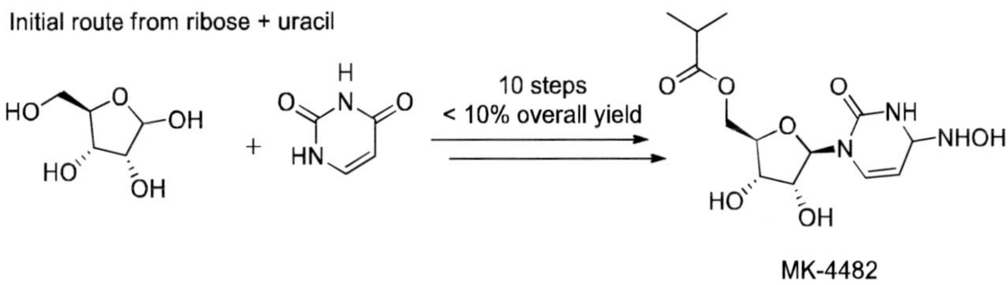

Scheme 1. The initial route to MK-4482 from Uridine.

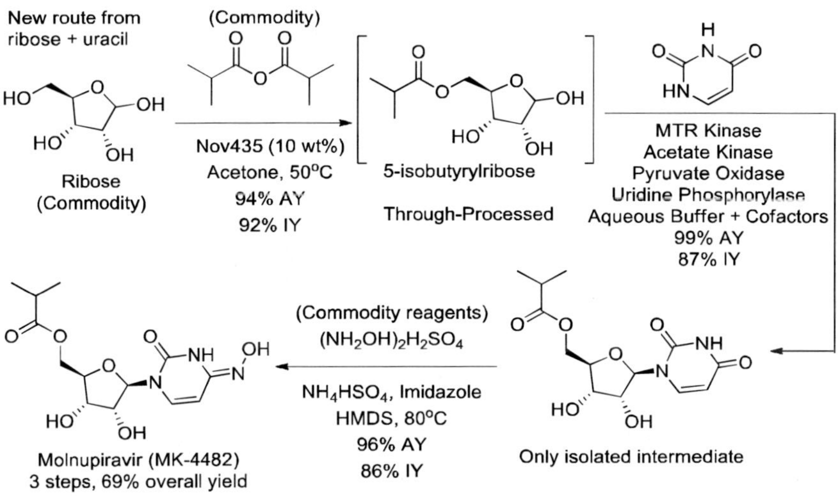

Scheme 2. Modified route to MK-4482 from Ribose.

A three-step synthetic methodology was reported by establishing the following advantages: (a) simple commodity reagents such as ribose, uracil, and hydroxylamine hydrogen sulfate; (b) use of a class 3 solvent, acetone; and (c) use of biocatalysis such as Novazyme 435 for selective esterification and MTR kinase, Acetate kinase, Pyruvate Oxidase, and Uridine Phosphorylase to carryout C-N bond formation (Scheme 2) [12, 13].

In a modified version, to reduce waste prevention, increasing atom economy and to decrease hazardous chemical synthesis, the numbers of steps were reduced to only six by using urdine as starting material in comparison with ribose and uracil employed methodology. Although overall yield of the process was increased to 17% in comparison with 10% by employing a nucleoside like Uridine as a starting material is found to be a better option in comparison with ribose, but still C-N bond formation requires few more additional steps (Scheme 3) [14].

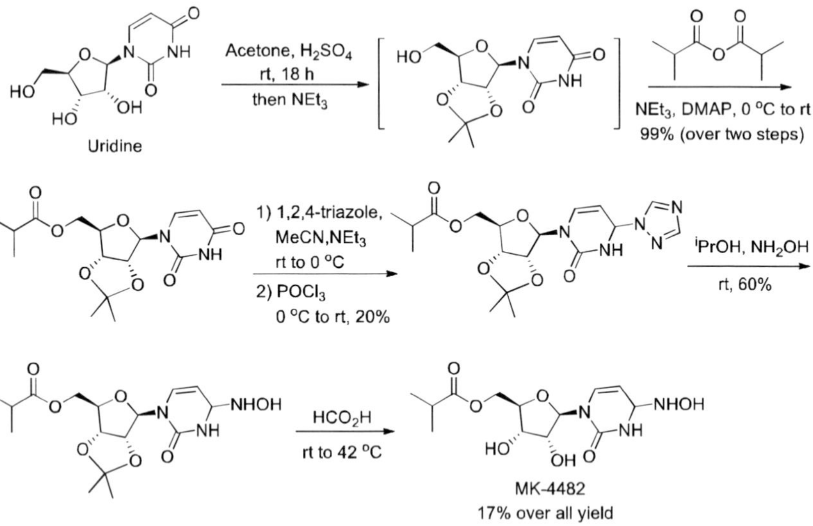

Scheme 3. MK-4482 from Uridine.

Further, the second-generation strategy was developed by Vasudevan et al. which was found to be a greener alternative to the synthesis route. By using cytidine instead of uridine, the number of steps were reduced to three, including the protection, esterification, and deprotection processes. All the reactions were optimized under milder conditions, and there were other interesting features, such as the catalytic process (DMAP-catalyzed esterification) (Scheme 4). Although in the third step a green solvent (class 3 solvent) like PrOH was employed to carry out the deprotection reaction, the usage of a class I solvent (i.e., a solvent that needs to be avoided) such as DMF as a reagent is the unavoidable solvent to carry out the protection of secondary hydroxyl groups and the amine of cytidine, which needs further development [15].

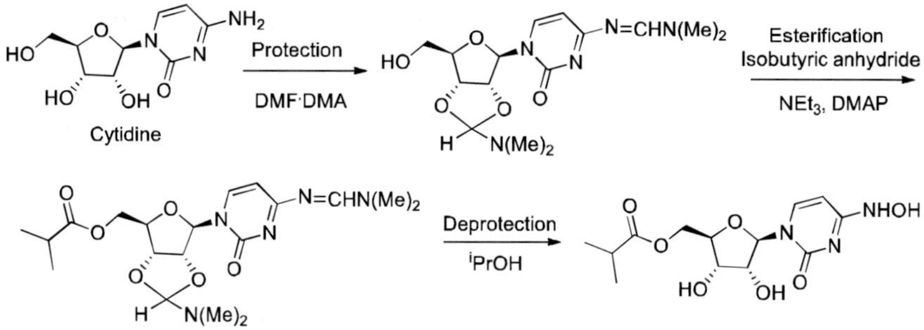

Scheme 4. MK-4482 from Cytidine.

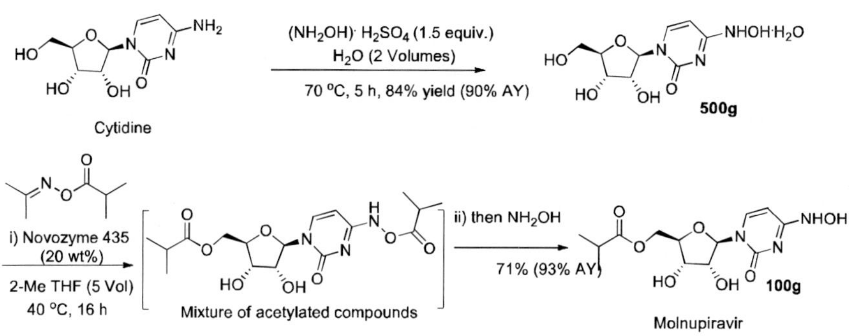

Scheme 5. MK 4482 from Cytidine.

In addition to the previous approaches, a green solvent-based two-step process was established under highly concentrated reaction conditions. The use of bio-renewable 2-MeTHF (7th principle consists of using renewable feedstocks) was considered as an additional greener technique that can be optimized, which was useful even for bulk scale (100 g). The execution of the 9th principle insists synthetic chemists to use catalyst in comparison with stoichiometric reagent, which was established during the selective esterification of cytidine using Novozyme 435 (20 wt.%). Another inherent advantage of this methodology is the use of biocatalysts as an alternative to chemical-based catalysis (Scheme 5) [16].

Paymode DJ et al. reported a higher (75%) overall yield by adopting the methodology employing higher equivalence of DMF-DMA and hydroxyl amine reagents, but usage of 1,4-dioxane as an alternative in comparison with a class-3 solvent, which was considered a limited category as a class -2 solvent during ester bond formation (Scheme 6). Out of the two optimized pathways, it was discovered that esterification is the first step, followed by hydroxamination, resulted in higher monupiravir yields [17].

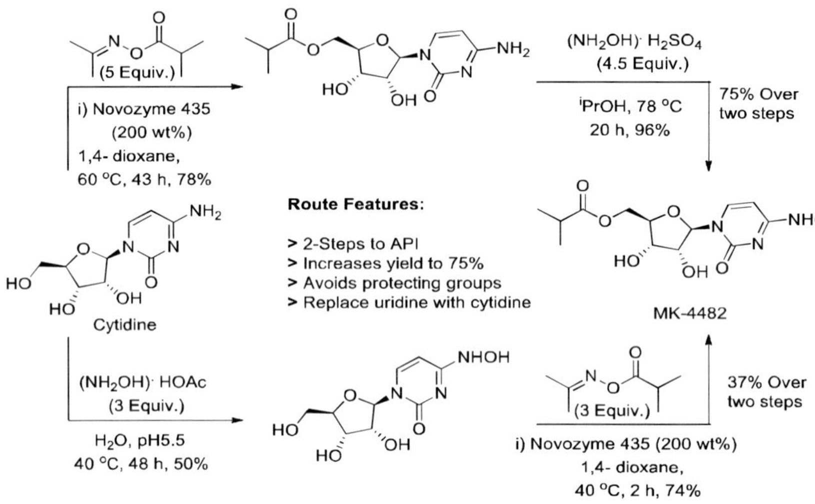

Scheme 6. A new route to MK-4482 from Cytidine.

One of the research groups recently reported an advanced two-step methodology for the synthesis of Molnupiravir using simple and commercially available reagents such as isobutyryl chloride and hydroxyl amine hydrogen sulphate under protection group-free, enzymatic-free (relatively cheaper) methodology (Scheme 7). The synthetic strategy starts with the cytidine nucleoside as substrate, and chemo selective acylation was optimized by using isobutyryl chloride as the acylating reagent and Et_3N as the base, which resulted in >89% of the product. Only the primary hydroxyl group of cytidines was acylated under highly diluted conditions (high dilution methodology) in DMF solvent (0.05 mmol/mL). The other achievement of this protocol relies on the hydroamination of the amine group of the cytidine ester using water as the solvent with a 96% yield. The overall yield of the process was found to be 87.5%. [18].

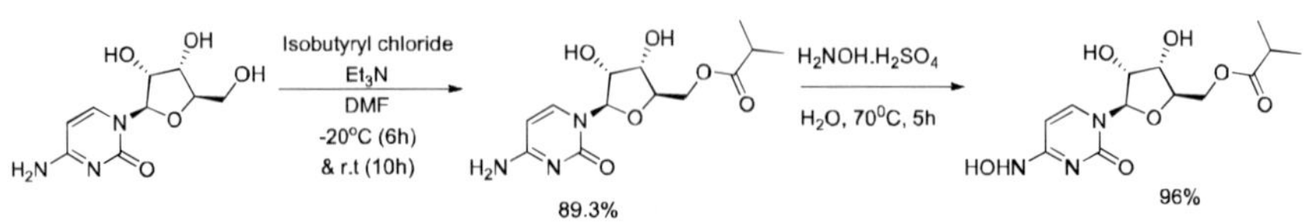

Scheme 7. Modified route to MK-4482 from Cytidine.

To achieve an ideal green synthetic route, major advancements were made by Hu et al. in 2022. Both esterification and hydroxyl amination were carried out by them using a one-pot process of synthesis route. For these purposes the following is necessary

- Organic solvents required to purify intermediates can be avoided, which satisfies the first green chemical principle.
- Energy and time consumption can be reduced, which satisfies the sixth green chemical principle (Scheme 8) [19].

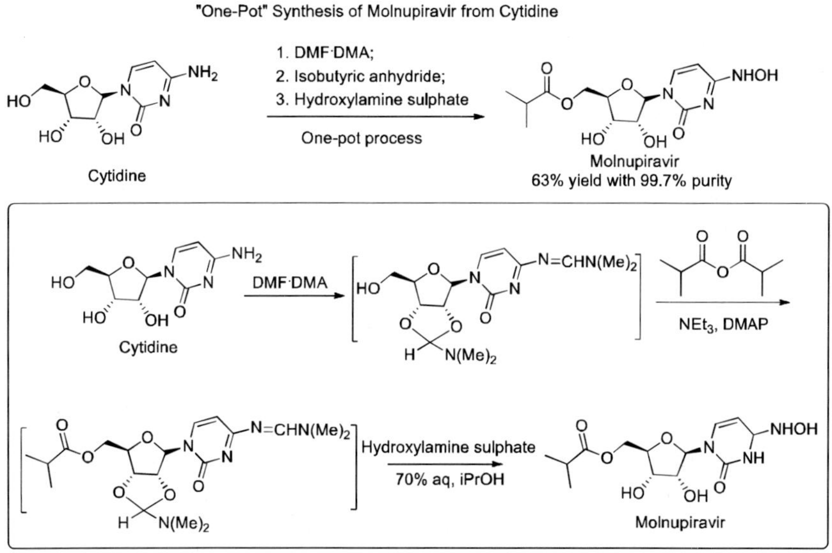

Scheme 8. One pot synthetic route to MK-4482.

Favipiravir

6-fluoro-3-hydroxypyrazine-2-carboxamide (Favipiravir, T-705, or AviganTM) is a nucleoside kind of prodrug that follows the mechanisms of both guanosine and adenosine. It was originally targeted to cure influenza (seasonal) by Toyama Chemical. Favipiravir (T-705) is proven to be a dependable medicine to treat influenza, which works selectively to inhibit the RNA-dependent RNA polymerase of influenza viruses (Furuta et al. 2013) [20]. Favipiravir follows a similar mechanism pathway to Remdesivir in that it actively participates in the inhibition of viral RNA-dependent RNA polymerase [21]. This molecule was initially examined to treat COVID-19 in Wuhan. Thereafter, numerous clinical trials have been conducted to assess its efficiency in the treatment of SARS-CoV-2. Although the data is inadequate, large-cohort clinical trials started in May 2021 and are presently continuing. Favipiravir is to be investigated as a possible COVID-19 treatment for at-home recovery in the principal trial. Pyrazine-based Favipiravir has three different substituents, such as F, OH, and $CONH_2$. This compound is acidic because of the hydroxyl group, and thereby it can show tautomerism to generate ketones. This product can be synthesized starting from different starting materials, such as 3-aminopyrazine-2-carboxylic acid, 3-hydroxypyrazine-2-carboxamide, 3-hydroxypyrazine-2-carboxylic acid, and pyrazin-2-amine (Figure 3) [22, 23].

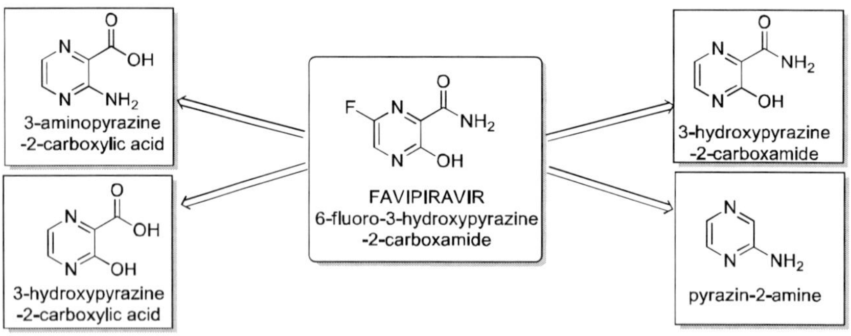

Figure 3. Retro synthetic approaches towards Favipiravir synthesis.

This biologically active product can be obtained using four different synthetic methodologies. Some researchers developed this original route by optimizing seven synthetic steps, which start with the commercially available 3-aminopyrazine-2-carboxylic acid. The overall yield of the process is only 0.44%. The first step involves esterification of the carboxylic acid functional group with MeOH in an acidic medium (H_2SO_4). In the second step, bromination was carried out on the methyl 3-aminopyrazine-2-carboxylate using NBS in CH_3CN solvent (Scheme 9) [24]. The third step includes both diazotizations followed by alcoholysis, which was carried out further in the MeOH solvent along with concentrated H_2SO_4 and the productivity is low (only 35%). The fourth step involves the formation of C-N bonds using a Pd catalyst. Even though this was a catalyst-based reaction, both the reagents $Pd_2(dba)_3$ and S-BINAP were discovered, which is found to be more expensive. Amide functional group was generated using ammonium hydroxide. Using Schiemann's methodology, fluorination was carried out with pyridinium hydrogen fluoride (Olah's reagent). Olah's reagent is highly corrosive, and hence this process is very difficult to execute for larger scale synthesis Finally, demethylation was executed using a combination of NaI and TMSCl, which resulted in Favipiravir, but the product yield for the last reaction is very poor (only 4.3% [25, 26].

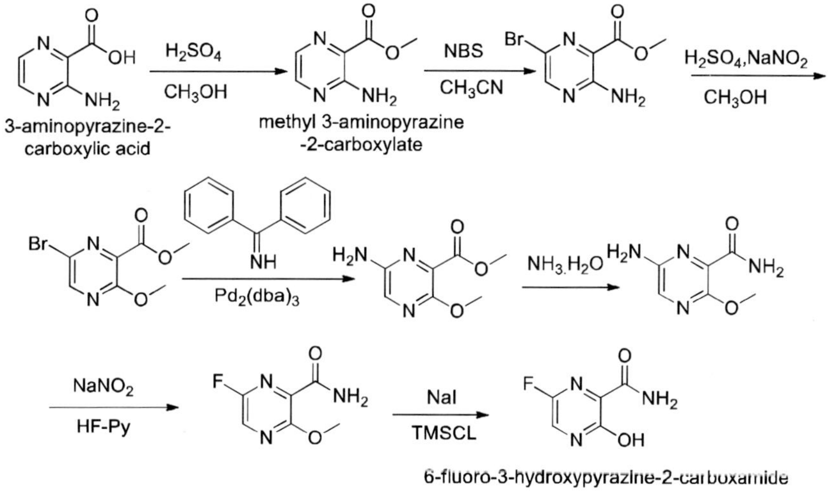

Scheme 9. Initial approach to Favipiravir.

A modified strategy with an overall yield of 8% was developed by relying solely on the four-step methodology reported by Beigel, et al. that begins with commercially available 3-hydroxypyrazine-2-carboxylic acid as starting material, which can be obtained from readily available 3-aminopyrazine-2-carboxylic acid (Scheme 10) [27]. The first step involves one-pot synthesis (78% yield) of amide by combining esterification and amide bond formation. Esterification was carried out using a combination of $SOCl_2$ and MeOH, whereas amide bond formation was executed with NH_4OH. The interesting thing about this methodology is that it relies on the execution of a chromatography free process, i.e., the recrystallization of the product with water. The second step involves the nitration process (48% yield), which was carried out by using con. H_2SO_4 and KNO_3, which is the major advantage in this step involves executing the nitration under an organic solvent-free reaction and purification process. The third step involves the reduction of the nitro functional group to the corresponding amine (77% yield) using Raney Ni and $NH_2NH_2H_2O$ in a greener MeOH solvent, and purification of the resultant product was not required to carry out Schiemann's fluorination using Olah's reagent (30% yield). Column chromatography purification is only required in the final step [28, 29].

Guo et al. reported a four-step methodology for the synthesis of Favipiravir starting from 3-hydroxypyrazine-2-carboxamide. The first step is nitration of 3-hydroxypyrazine-2-carboxamide by using a combination of KNO_3 and H_2SO_4 (low yield, 18% for the nitration), followed by chlorination as the second step using $POCl_3$ along with pyridine. Fluorination was carried out in the subsequent step using KF along with a Bu_4NBr catalyst in DMSO. Hydration of the nitrile functional group was carried out as an aqueous phase transformation using H_2O_2 in an alkaline solution. Finally, regioselective fluorination in water solvent was accomplished using $NaHCO_3$ (Scheme 11) [30].

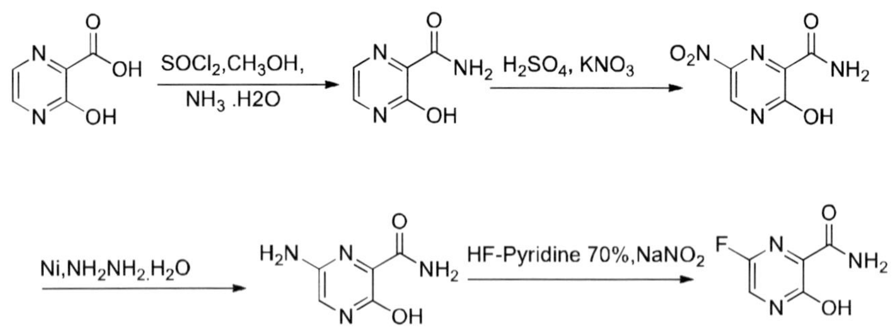

Scheme 10. Synthesis of Favipiravir from 3-hydroxypyrazine-2-carboxylic acid.

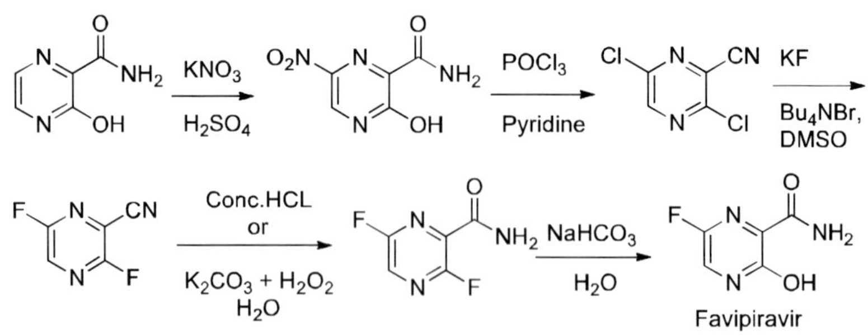

Scheme 11. Synthesis of Favipiravir from 3-hydroxypyrazine-2-carboxamide.

To overcome these drawbacks such as (i) lower overall yield and (ii) costlier starting material, Hara et al. modified this protocol further through the preparation of a common intermediate. Commercially cheaper, 3-aminopyrazine-2-carboxylic acid was used as the starting material, and it was esterified using a CH_3OH along with H_2SO_4. Bromination of the product was carried out using NBS in CH_3CN (Scheme 12). The third step includes both diazotizations followed by hydrolysis using H_2SO_4 and H_2O. Using NH_4OH, methyl ester as the functional group was transformed into the required amide group. Finally, using $POCl_3$ and DIEA, the common intermediate 3,6-dichloropyrazine-2-carbonitrile was obtained, which was shown in the scheme of synthesized pathway and appears until this common intermediate, the overall yield was 37%. Based on green chemical principles, this route is not preferable due to the higher number of reactions (seven) in comparison with that of the scheme, which consists of only $POCl_3$, which is highly toxic and corrosive in nature. Since large amounts of reagents are required, that considered as an important disadvantage in these two schemes during bulk scale processes [31, 32].

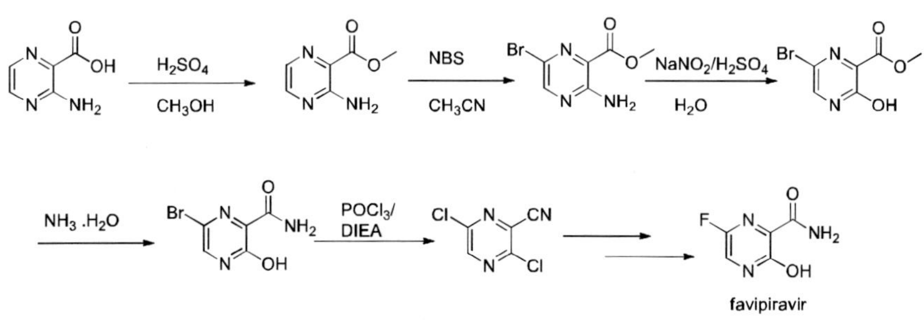

Scheme 12. Synthesis of Favipiravir from 3-hydroxypyrazine-2- carboxylic acid.

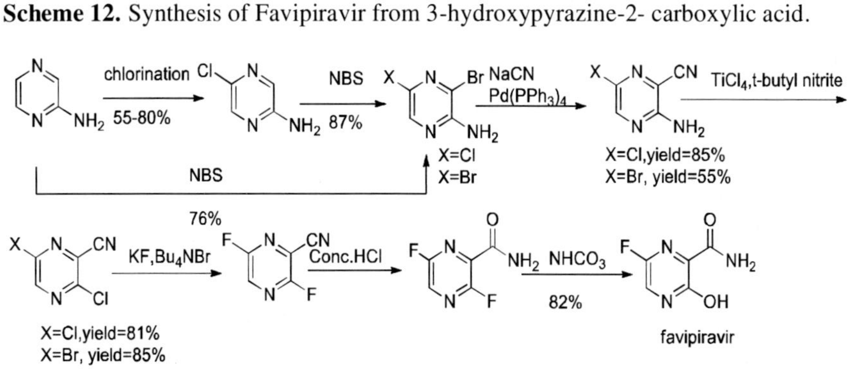

Scheme 13. Synthesis of Favipiravir from 3-hydroxypyrazine-2- carboxylic acid.

To address these issues, Qi et al. reported an alternative methodology to avoid $POCl_3$. The required common intermediate was prepared in 4 steps with a higher overall yield of up to 48%. To avoid the use of $POCl_3$, commercially available, cheaper pyrazin-2-amine was used as the starting material, and mono chlorination reaction was carried out using N-Chloro-N-methoxytoluenesulfonamide as the chlorinating agent, where the yield of about 80% was reported [33] (Scheme 13).

Now, this monochlorinated product was utilized to carry out regioselective bromination for the synthesis of a bromo-substituted product using NBS (87%). Where NaCN was used to install the nitrile functional group by the chemo-selective substitution (85% yield), which was catalyzed by $Pd(PPh_3)_4$. The Sandmeyer reaction was further executed using a combination of $TiCl_4$ and t-butyl nitrite to synthesize the common intermediate 3,6-dichloropyrazine-2-carbonitrile. The use of highly toxic NaCN was a potential drawback in this procedure, which should be replaced with other reagents in the future to achieve green pathway of synthesis [34].

Remdesivir

Remdesivir, which consists of two parts, the ribose core, and the unnatural base, which is found to be one of the promising COVID-19 drugs that has been permitted by the FDA to treat patients with COVID-19. This drug has played a crucial role in treating patients infected with the SARS-CoV-2 virus. Even though there are minute side effects that have been attributed to this drug, it is very important to treat the SARS-CoV-2 virus by application of this drug. Remdesivir was highly effective in treating the severe acute respiratory issues caused by the COVID-19 pandemics. Initially, Hepatitis C and Ebola virus infections were also effectively treated with this drug developed by Gilead Sciences. [30, 31] Syndrome Corona virus 2 has been affecting a nation's health and economy all over the world [32-37]· To overcome this pandemic situation, there is an urgent need of highly valuable and affordable treatments that are still to be investigated. The key mechanism that occurs for viral replication due to the RNA-dependent RNA polymerase (RdRp) enzyme of the SARS-CoV-2 virus can be effectively stopped by this drug. Figure 4 depicts the available routes for remdesivir synthesis using GS-441524, ribose, and modified nucleobase [35, 36].

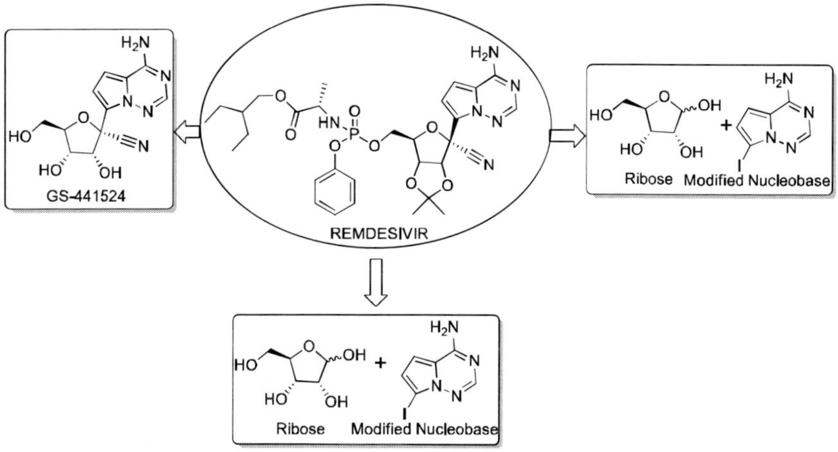

Figure 4. Retro synthetic approaches towards the preparation of Remdesivir.

A novel synthetic route developed towards one of the most efficient drugs to treat COVID-19, (remdesivir) was first developed by Warren et al. [36]. Initially, it took 9 to 12 months for the back-to-back manufacturing process of this drug, but later time this was reduced to six months in January 2020. Gilead Sciences was successful in finding the ways to increase production of this valuable drugs in a shorter period. Protecting group manipulations and increasing the number of steps reduced at the time of synthesis, the overall yield to less than 10%. The use of a greater number of chemicals results somewhat expensive, and low-yield production steps make this route difficult for a sufficient production and supply of remdesivir drug (Scheme 14) [37, 38].

The improved yields for preparing remdesivir were achieved by using halogenated unnatural bases, i.e., iodinated unnatural bases, instead of brominated unnatural bases, which is the key step of glycosylation, making this route better than the initial route developed by Gilead Sciences. The use of class -3 solvents such as acetone makes this process more environmentally friendly. The use of the milder reagent i-PrMgCl.LiCl instead of large amounts of the strong base n-BuLi allowed for easy metal halogen exchange and the execution of large-scale batches with consistent yields (Scheme 15). The overall usage of less hazardous chemicals is one of the added advantages of this route. Even though it improved the yield, the more number of steps and yield improvement, up to a certain extent, created ample opportunities for further development of the remdesivir preparation [39].

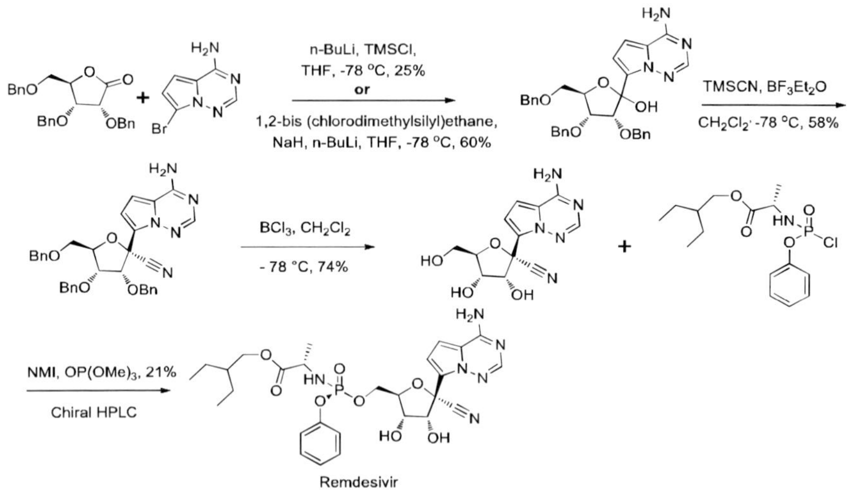

Scheme 14. First generation Gilead Sciences route to Remdesivir from Ribose.

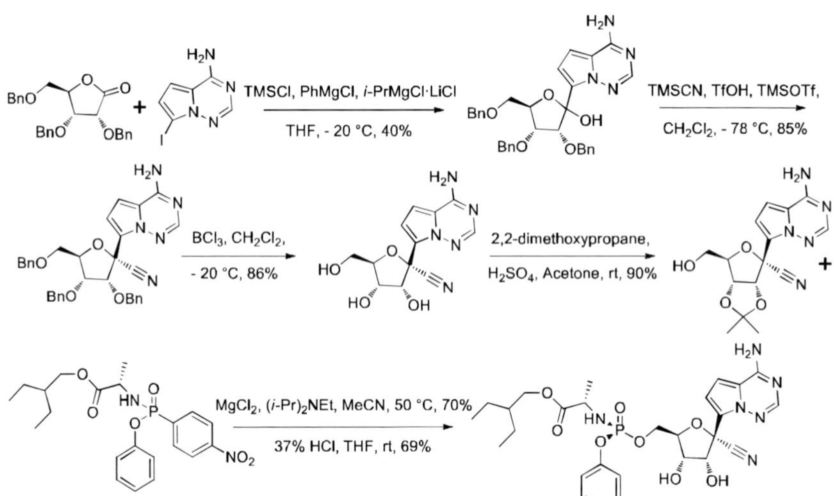

Scheme 15. Second generation Gilead Sciences route to Remdesivir from Ribose.

Zhang's group reported an excellent stereo selective synthesis route by following the 8th principle of green chemistry, which requires synthetic chemists to reduce the number of steps. Chiral imidazole derivatives were used to catalyse a pivotal step in the formation of nucleotide monophosphate compound with an excellent overall yield of 73%, reducing the use of a greater number of chemicals in the separate preparation of enantiomerically pure phosphorylated agent (Scheme 16) [40].

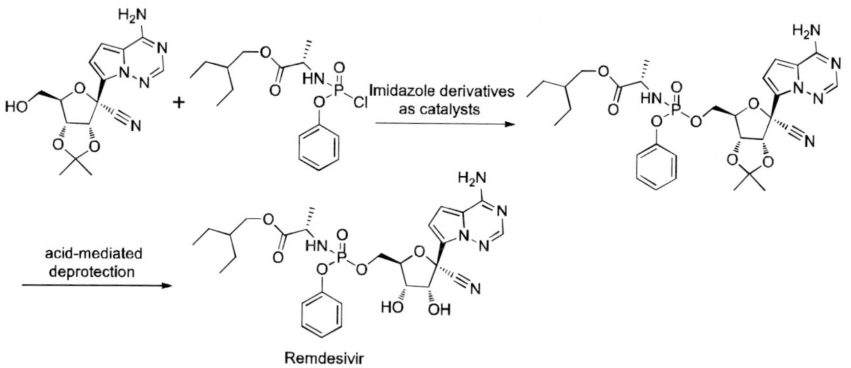

Scheme 16. Imidazole derivatives catalysis route to Remdesivir from acetal protected Ribose.

The use of simple chemicals for the masking and unmasking of some of the functional groups in organic chemistry makes the synthesis route valuable instead of using hazardous chemicals, which cause environmental pollution. A facile and new route has been established to access remidesivir by avoiding the masking and unmasking of some functional groups with hazardous chemicals, even with excellent yields (Scheme 17). The four major advantages of this route are as follows:

- BCl_3 is one of the hazardous chemicals used for benzyl deprotection, which can be avoided by applying the third green chemistry principle.
- The key coupling reaction in the preparation of remdesivir was successfully obtained without the addition of any toxic chemicals.
- Allyl group protection and deprotection could be achieved by using suitable catalytic methods.
- The key coupling reaction in the preparation of remdesivir yield increased by up to 25%, making this method comparatively more suitable for preparing remdesivir drug on a large scale, and it is also used to synthesize novel anti-viral analogues [41, 42].

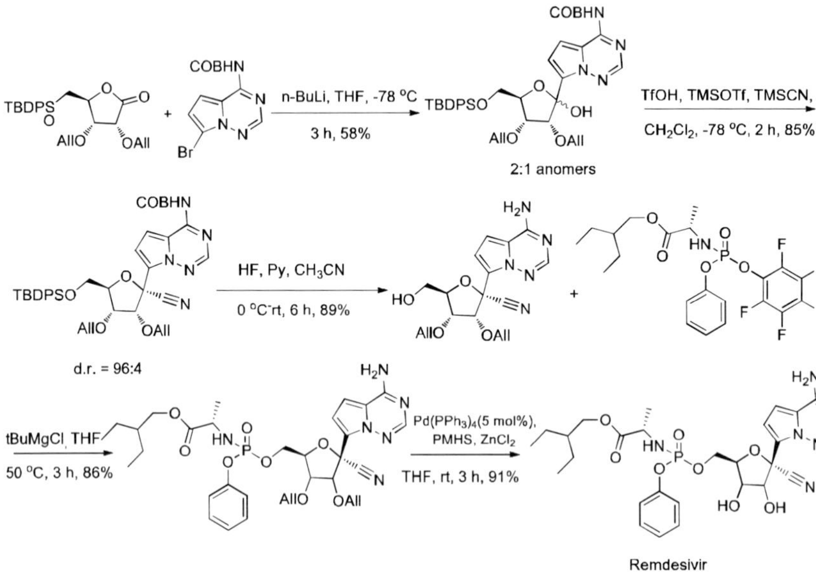

Scheme 17. Grignard reaction based Remdesivir synthesis.

Masking and unmasking of some of the functional groups in organic chemistry require a large number of steps and the use of numerous chemicals in the process of synthesis reduces the efficiency of the reactions. Very recently, starting from GS-441524, a remarkable short sequence was achieved in the preparation of remdesivir using protecting group *N, N-dimethylformamide* dimethyl acetal (Scheme 18). It involves only three steps (protection, phosphoramidation, and deprotection) [43]. The execution of this approach has three major advantages, which includes,

- Execution of the first green chemistry principle to prevent waste generation, which is attained by avoiding generation of the degraded impurity by employing milder reaction conditions.
- A phosphorylated reaction was achieved with high stereoselectivity to minimize wastes, which is also another value added to the green chemistry approach.
- Column chromatography was applied successfully for intermediate purification, which is one of the significant applications of the 5th green chemistry principle to reduce solvent usage. In summary, according to all reports, this method revealed an excellent pathway to

prepare remdesivir from GS-441524 in a highly efficient synthesis route.

- It is also observed to be the best method for achieving 85% overall yield through a three-step sequence, as compared with the other methods, where 43% overall yield was reported in the literature. In addition to achieving high yields, this method avoided impurity formation and can also avoided intermediate purification resulting an appropriate value to green chemistry [44, 45].

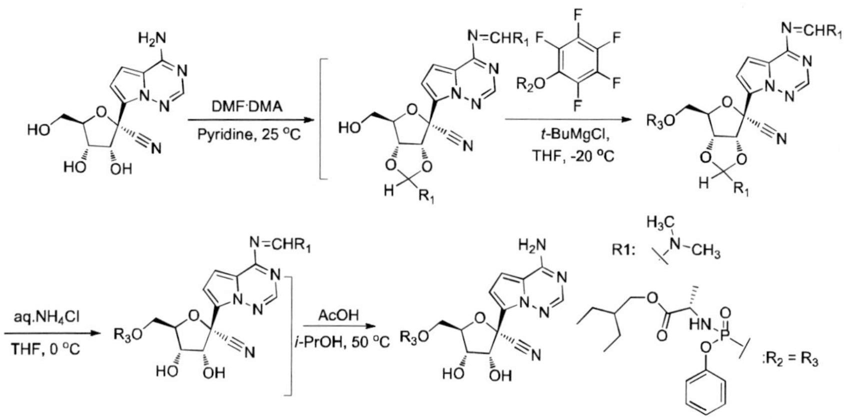

Scheme 18. DMF.DMA based Remdesivir synthesis.

Conclusion

At present, these synthetic methodologies were developed to meet the current demand of the drugs belongs to antiviral category and win over this unavoidable battle with COVID-19 variants. In the case of molnupiravir, one-pot synthetic methodologies in addition to enzymatic reactions were developed, which was found to be highly effective. But still the appropriate drug for this dangerous virous (COVID-19, *Corona vinous*) is not developed. The various categories of drugs used for the treatment of corona vinous is although partially successful, but still the proper drug is not developed worldwide. The pathway of the synthesis of this drug required a variety of toxic chemical, therefore the green methodologies of synthesis of these drugs such as remdesivir, favipiravir, molnufiravir in cost effective way is highly essential and challenge to the human society in the present situation to prevent

COVID 19. The path of green synthesis can be achieved by using low-boiling solvents such as DCM and THF, which must have to be planned by the organic chemists in this regard. Over the last two decades, enormous environmentally friendly green chemistry approaches have been evaluated in drug preparation. Many enzymatic and nonenzymatic green synthetic protocols were developed to prepare the above-mentioned drugs. But the development of appropriate drug for treatment of COVID-19 and the green pathway of the synthesis of the drugs used presently for prevention and treatment of this virous is basically a challenge to the global community to save the human community. Some important protocols, particularly for COVID-19 drugs, were discussed in detail in this chapter.

References

[1] World Health Organization. (2020). *Coronavirus disease* (COVID-19), 12 October 2020.

[2] Serafin, M. B., Bottega, A., Foletto, V. S., da Rosa, T. F., Hörner, A., & Hörner, R. (2020). Drug repositioning is an alternative for the treatment of coronavirus COVID-19. *International Journal of Antimicrobial Agents*, *55*(6), 1-4.

[3] Zheng, W., Sun, W., & Simeonov, A. (2018). Drug repurposing screens and synergistic drug-combinations for infectious diseases. *British journal of pharmacology*, *175*(2), 181-191.

[4] Zheng, W., Sun, W., & Simeonov, A. (2018). Drug repurposing screens and synergistic drug-combinations for infectious diseases. *British journal of pharmacology*, *175*(2), 181-191.

[5] Fischer, W. A., II Jr, E. J., Holman, W., Cohen, M. S., Fang, L., Szewczyk, L. J., ... & Painter, W. P. (2022). A phase 2a clinical trial of molnupiravir in patients with COVID-19 shows accelerated SARS-CoV-2 RNA clearance and elimination of infectious virus. *Sci Transl Med.* 14(628), 1-11.

[6] Wu, R., Wang, L., Kuo, H. C. D., Shannar, A., Peter, R., Chou, P. J., & Kong, A. N. (2020). An update on current therapeutic drugs treating COVID-19. *Current pharmacology reports*, *6*, 56-70.

[7] Singh, A. K., Singh, A., Singh, R., & Misra, A. (2021). Molnupiravir in COVID-19: A systematic review of literature. *Diabetes & Metabolic Syndrome: Clinical Research & Reviews*, *15*(6), 1-12.

[8] Heustess, A. M., Allard, M. A., Thompson, D. K., & Fasinu, P. S. (2021). Clinical management of COVID-19: a review of pharmacological treatment options. *Pharmaceuticals*, *14*(6), 520-548.

[9] Rizk, J. G., Forthal, D. N., Kalantar-Zadeh, K., Mehra, M. R., Lavie, C. J., Rizk, Y.,& Lewin, J. C. (2021). Expanded Access Programs, compassionate drug use, and

Emergency Use Authorizations during the COVID-19 pandemic. *Drug discovery today*, *26*(2), 593-603.

[10] Benkovics, T., McIntosh, J., Silverman, S., Kong, J., Maligres, P., Itoh, T., & Fier, P. (2020). *Evolving to an ideal synthesis of molnupiravir, an investigational treatment for COVID-19.*1-5. https://doi.org/10.26434/chemrxiv.13472373.v1.

[11] Scavone, C., Mascolo, A., Rafaniello, C., Sportiello, L., Trama, U., Zoccoli, A., & Capuano, A. (2022). Therapeutic strategies to fight COVID-19: Which is the status artis? *British Journal of Pharmacology*, *179*(10), 2128-2148.

[12] McIntosh, J. A., Benkovics, T., Silverman, S. M., Huffman, M. A., Kong, J., Maligres, P. E., & Fier, P. S. (2021). Engineered ribosyl-1-kinase enables concise synthesis of molnupiravir, an antiviral for COVID-19. *ACS Central Science*, *7*(12), 1980-1985.

[13] Gopalsamuthiram, V., Williams, C., Noble, J., Jamison, T. F., Gupton, B. F., & Snead, D. R. (2021). A Concise Route to MK-4482 (EIDD-2801) from Cytidine: Part 2. *Synlett*, *32*(03), 326-328.

[14] McIntosh, JA, Benkovics, T, Silverman, SM, Huffman, MA, Kong, J, Maligres, PE, (2021) Engineered Ribosyl-1-Kinase Enables Concise Synthesis of Molnupiravir, an Antiviral for COVID-19. *ACS Cent Sci.* 7(12),1980-1985.

[15] Vasudevan, N., Ahlqvist, G. P., McGeough, C. P., Paymode, D. J., Cardoso, F. S., Lucas, T., & Snead, D. R. (2020). A concise route to MK-4482 (EIDD-2801) from cytidine. *Chemical Communications*, *56*(87), 13363-13364.

[16] Hu, T., Xie, Y., Zhu, F., Gong, X., Liu, Y., Xue, H., & Shen, J. (2022). "One-Pot" Synthesis of Molnupiravir from Cytidine. *Organic Process Research & Development*, *26*(2), 358-364.

[17] Paymode, D. J., Vasudevan, N., Ahmad, S., Kadam, A. L., Cardoso, F. S., Burns, J. M., ... & Snead, D. R. (2021). Toward a practical, two-step process for molnupiravir: Direct hydroxamination of cytidine followed by selective esterification. *Organic Process Research & Development*, *25*(8), 1822-1830.

[18] Venkatanarayana, P., Kolli, D., Seelama, N. V., & Ramakrishna, D. S. (2022). Synthesis of molnupiravir (MK-4482, EIDD-2801): a promising oral drug for the treatment of COVID-19 starting from cytidine. *Nucleosides, Nucleotides & Nucleic Acids*, 1-9. https://doi.org/10.1080/15257770.2022.2153140.

[19] Hu, T., Xie, Y., Zhu, F., Gong, X., Liu, Y., Xue, H., & Shen, J. (2022). "One-Pot" Synthesis of Molnupiravir from Cytidine. *Organic Process Research & Development*, *26*(2), 358-364.

[20] Furuta, Y., Gowen, B. B., Takahashi, K., Shiraki, K., Smee, D. F., & Barnard, D. L. (2013). Favipiravir (T-705), a novel viral RNA polymerase inhibitor. *Antiviral research*, *100*(2), 446-454.

[21] Joshi, S., Parkar, J., Ansari, A., Vora, A., Talwar, D., Tiwaskar, M., & Barkate, H. (2021). Role of favipiravir in the treatment of COVID-19. *International Journal of Infectious Diseases*, *102*, 501-508.

[22] Baranovich, T., Wong, S. S., Armstrong, J., Marjuki, H., Webby, R. J., Webster, R. G., & Govorkova, E. A. (2013). T-705 (favipiravir) induces lethal mutagenesis in influenza A H1N1 viruses in vitro. *Journal of virology*, *87*(7), 3741-3751.

[23] Oestereich, L., Lüdtke, A., Wurr, S., Rieger, T., Muñoz-Fontela, C., & Günther, S. (2014). Successful treatment of advanced Ebola virus infection with T-705 (favipiravir) in a small animal model. *Antiviral research*, *105*, 17-21.

[24] Furuta, Y., Komeno, T., & Nakamura, T. (2017). Favipiravir (T-705), a broad-spectrum inhibitor of viral RNA polymerase. *Proceedings of the Japan Academy, Series B*, *93*(7), 449-463.

[25] Jin, Z., Smith, L. K., Rajwanshi, V. K., Kim, B., & Deval, J. (2013). The ambiguous base-pairing and high substrate efficiency of T-705 (favipiravir) ribofuranosyl 5′-triphosphate towards influenza A virus polymerase. *PloS one*, *8*(7), 1-10.

[26] Furuta, Y., Gowen, B. B., Takahashi, K., Shiraki, K., Smee, D. F., & Barnard, D. L. (2013). Favipiravir (T-705), a novel viral RNA polymerase inhibitor. *Antiviral research*, *100*(2), 446-454.

[27] Beigel, J. H., Tomashek, K. M., Dodd, L. E., Mehta, A. K., Zingman, B. S., Kalil, A. C., ... & Lane, H. C. (2020). Remdesivir for the treatment of Covid-19. *New England Journal of Medicine*, *383*(19), 1813-1826.

[28] Guo, Q., Xu, M., Guo, S., Zhu, F., Xie, Y., & Shen, J. (2019). The complete synthesis of favipiravir from 2-aminopyrazine. *Chemical Papers*, *73*, 1043-1051.

[29] Karatas, H., Hanashalshahaby, E. H. A., Catal, U., Butun, Y. E., Kurt, E., Gursel, S., ... & Guzel, M. (2022). An economical and practical procedure of favipiravir synthesis for the treatment of Covid-19. *Chemical Papers*, 1-8.

[30] El-Nahas, A., Hirao, K. A (1999) Theoretical study on 2-hydroxypyrazine and 2,3-dihydroxypyrazine: tautomerism, intramolecular hydrogen bond, solvent effects. *Journal of Molecular Structure: THEOCHEM*. 459 (1-3), 229-237.

[31] Hara, T., Norimatsu, N., Kurushima, H., & Kano, T. (2014). U.S. Patent No. 8,835,636. Washington, DC: U.S. Patent and Trademark Office.

[32] Uddin, E., Islam, R., Ashrafuzzaman, Bitu, N. A., Hossain, M. S., Islam, A. N., Asraf, A., Hossen, F., Mohapatra, R. K., & Kudrat-E-Zahan, M. (2021). Potential Drugs for the Treatment of COVID-19: Synthesis, Brief History and Application. *Current drug research reviews*, *13*(3), 184–202.

[33] Qi, Q., Zhuang, L., Shen, Y., Geng, Y., Yu, S., Chen, H., & Chen, Z. (2016). A novel systemic inflammation response index (SIRI) for predicting the survival of patients with pancreatic cancer after chemotherapy. *Cancer*, 122(14), 2158-2167.

[34] Tomar, R., Sahni, A., Chandra, I., Tomar, V., & Chandra, R. (2018). Review of noscapine and its analogues as potential anti-cancer drugs. *Mini-Reviews in Organic Chemistry*, *15*(5), 345-363.

[35] Wang, M., Zhang, L., Huo, X., Zhang, Z., Yuan, Q., Li, P.,& Zhang, W. (2020). Catalytic asymmetric synthesis of the anti-COVID-19 drugs remdesivir. *Angewandte Chemie International Edition*, *59*(47), 20814-20819.

[36] Warren, T. K., Jordan, R., Lo, M. K., Ray, A. S., Mackman, R. L., Soloveva, V., ... & Bavari, S. (2016). Therapeutic efficacy of the small molecule GS-5734 against Ebola virus in rhesus monkeys. *Nature*, *531*(7594), 381-385.

[37] Wang, M., Zhang, L., Huo, X., Zhang, Z., Yuan, Q., Li, P., & Zhang, W. (2020). Catalytic asymmetric synthesis of the anti-COVID-19 drugs remdesivir. *Angewandte Chemie International Edition*, *59*(47), 20814-20819.

[38] Palli, K. K., Ghosh, P., Avula, S. K., Rao, B. S. S., Patil, A. D., Ghosh, S., & Chandrasekhar, S. (2022). Total synthesis of remdesivir. *Tetrahedron Letters*, *88*, 1-4.

[39] Reeves, J. T., Fandrick, D. R., Tan, Z., Song, J. J., Rodriguez, S., Qu, B., & Senanayake, C. H. (2013). Development of a large scale asymmetric synthesis of the glucocorticoid agonist BI 653048 BS H3PO4. *The Journal of Organic Chemistry*, *78*(8), 3616-3635.

[40] Fan, Q., Zhang, B., Ma, J., & Zhang, S. (2020). Safety profile of the antiviral drug remdesivir: An update. *Biomedicine & Pharmacotherapy*, *130*, 1-4.

[41] Palli, K. K., Ghosh, P., Avula, S. K., Rao, B. S. S., Patil, A. D., Ghosh, S., & Chandrasekhar, S. (2022). Total synthesis of remdesivir. *Tetrahedron Letters*, *88*, 153590. (1-4).

[42] Khanal, P. (2020). Remdesivir for COVID-19 treatment: mechanism OF action, synthesis, and clinical trials. *J Pharm Pharm Sci*, *9*, 1062-1068.

[43] Alvarez, J. C., Moine, P., Etting, I., Annane, D., & Larabi, I. A. (2020). Quantification of plasma remdesivir and its metabolite GS-441524 using liquid chromatography coupled to tandem mass spectrometry. Application to a Covid-19 treated patient. *Clinical Chemistry and Laboratory Medicine (CCLM)*, *58*(9), 1461-1468.

[44] Humeniuk, R., Mathias, A., Cao, H., Osinusi, A., Shen, G., Chng, E., & German, P. (2020). Safety, tolerability, and pharmacokinetics of remdesivir, an antiviral for treatment of COVID-19, in healthy subjects. *Clinical and translational science*, *13*(5), 896-906.

[45] Li, Y., Cao, L., Li, G., Cong, F., Li, Y., Sun, J., ... & Zhang, X. (2021). Remdesivir metabolite GS-441524 effectively inhibits SARS-CoV-2 infection in mouse models. *Journal of medicinal chemistry*, *65*(4), 2785-2793.

Index

T

Z